Amazon® Fire TV

by Paul McFedries

Amazon® Fire TV For Dummies®

Published by: **John Wiley & Sons, Inc.,** 111 River Street, Hoboken, NJ 07030-5774, www.wiley.com

Copyright © 2020 by John Wiley & Sons, Inc., Hoboken, New Jersey

Published simultaneously in Canada

Contents at a Glance

Table of Contents

Introduction

Many of the gadgets we use every day don't look all that sophisticated from the outside, but they're brimming with hidden features and options. Think of the average smartphone, which to the uninitiated (if such people still exist) looks like a shiny block of glass and metal, but is really a computer more powerful than the supercomputers of yesteryear. Think of Alexa, Amazon's voice assistant, which seems like the very definition of simplicity — "Alexa, bark like a dog" — but can do some truly amazing things (as I explain in my book *Alexa For Dummies* [Wiley]).

The pile of deceptively simple electronics also includes Amazon's Fire TV devices. The most basic of these devices — the Fire TV Stick and Fire TV Stick 4K — look like nothing more than oversized thumb drives, while the Fire TV Cube looks like it could play the role of an airplane's black box flight recorder in a TV movie. But within these nondescript exteriors lie some sophisticated hardware and software that can do some amazing — and surprising — things.

So, yep, the basics of Fire TV are readily mastered, but to get at the hidden depths and power of Fire TV, you need a guide. *Amazon Fire TV For Dummies* aims to be that guide.

About This Book

This book takes you on a complete tour of Fire TV's capabilities, features, tools, and settings. In this book, you find everything you need to know to get the most out of your Fire TV investment.

Amazon Fire TV For Dummies boasts 12 chapters, but you don't have to read them from beginning to end — you can start wherever you want. Use the table of contents or index to find the information you need, and dip into the book when you have a question about Fire TV.

If your time is very limited (or you're just aching to start bingeing that new hot show), you can also ignore anything marked by the Technical Stuff icon or the information in sidebars (the gray-shaded boxes). Yes, these tidbits are fascinating (if I do say so myself), but they aren't critical to the subject at hand, so you won't miss anything critical by skipping them.

Finally, within this book, you may note that some web addresses break across two lines of text. If you're reading this book in print and want to visit one of these web pages, key in the web address exactly as it's written in the text, pretending as though the line break doesn't exist. If you're reading this as an e-book, you've got it easy — just click the web address to be taken directly to the web page.

Foolish Assumptions

This book is for people who are new (or relatively new) to using the Fire TV media-streaming device. Therefore, I do *not* assume that you're a Fire TV expert, a Fire TV connoisseur, or a Fire TV authority. However, I do assume the following:

>> You have a TV that's compatible with Fire TV (see Chapter 2).

>> You know how to connect devices to that TV.

>> You have a running Wi-Fi network with an Internet connection.

>> You know the password for your Wi-Fi network.

>> You have either an iOS or an Android mobile device (that is, a smartphone or tablet).

>> You know how to install and operate apps on your mobile device.

Icons Used in This Book

Like other books in the *For Dummies* series, this book uses icons, or little pictures in the margin, to draw your attention to certain kinds of material. Here are the icons that I use:

REMEMBER

Whenever I tell you something useful or important enough that you'd do well to store the information somewhere safe in your memory for later recall, I flag it with the Remember icon.

TECHNICAL STUFF

The Technical Stuff marks text that contains some for-nerds-only technical details or explanations that you're free to skip.

TIP

The Tip icon marks shortcuts or easier ways to do things, which I hope will make your life — or, at least, the Fire TV portion of your life — more efficient.

WARNING

The Warning icon marks text that contains a friendly but unusually insistent reminder to avoid doing something. You have been warned.

Beyond the Book

In addition to what you're reading right now, this product comes with a free access-anywhere Cheat Sheet that includes the most useful Fire TV settings, an exhaustive list of Alexa voice commands for controlling Fire TV, and a glossary of Fire TV terms to drop casually at your next cocktail party. To get this Cheat Sheet, go to www.dummies.com and type **Amazon Fire TV For Dummies Cheat Sheet** in the Search box.

Where to Go from Here

If you've had Fire TV for a while and you're familiar with the basics, you can probably get away with skipping the first five chapters and then dive into any part of the book that tickles your curiosity bone.

However, if you and Fire TV haven't met yet — particularly if you're not even sure what Fire TV *does* — this book has got you covered. To get your relationship with Fire TV off to fine start, I highly recommend that you read the book's first two chapters to get some of the basics down cold. Then read Chapter 3, 4, or 5, depending on which Fire TV device you've got. From there, you can head anywhere you like, safe in the knowledge that you've got some survival skills to fall back on!

1

Getting Started

IN THIS PART . . .

Find out what Fire TV is, what Fire TV can do, and what hardware you need to use Fire TV.

Welcome Fire TV into your home by learning where to put your Fire TV device, getting your device on your network, and customizing Fire TV.

Discover some crucial basics about your Fire TV device.

Chapter **1**

Understanding Streaming Media

Electronics such as Amazon's Fire TV devices — the Fire TV Stick, Fire Stick 4K, and Fire TV Cube — are streaming media devices.

If that sentence makes perfect sense to you, then you may want to move on to Chapter 2 because the next few pages probably won't tell you anything you don't already know. However, if you're scratching your head in bafflement, get comfy and prepare to be de-baffled!

In this chapter, you explore the two-sides-of-the-same-coin ideas of *streaming media* and *streaming media devices,* which are at the core of the Fire TV experience. Do you really need to know this background to use Fire TV? Will having a working definition of *streaming* benefit you when you're binge-watching *Fleabag?* Well, okay, the honest answer is "No" on both counts. Or, I should say, the answer is "No" *if* you don't care about getting the most out of your Fire TV investment, you're not the least bit curious how this technology works, or you're 100 percent certain that you'll never have problems with Fire TV. If that's you, start flipping ahead to the next chapter; otherwise, it's time to discover what this streaming business is all about.

Introducing Streaming

If you were around for the early days of the web — I'm talking about the mid to late 1990s — then you probably remember when the web pages of that era, which contained mostly text with a few images, started giving way to pages that contained *media* — music, TV shows, and even short movies. That was a fun development, but no one would have described it as "on-demand" entertainment because it could take anywhere from a few minutes to an hour or more for the media you clicked to download to your computer. Crucially, you had to wait until the entire media file was downloaded before you could start the playback. Inevitably, with Murphy's Law ("Anything that can go wrong will go wrong") in full effect back then (as it is today), the longer you had to wait for a media file to download, the more likely it was that the download would crash when it was 99 percent complete.

The molasses-in-January pace and the don't-breathe-until-it's-done fragility of media downloads were facts of online life back then, but a few nerds started thinking there had to be a better way. They realized that for most people, however slow their download speed, it was still faster than the rate at which they listened to or watched whatever media was being downloaded. Why was this speed difference important? Because it meant that after at least *some* of the media was downloaded, it could start playing from the beginning and the rest of the download could continue in the background without fear that the user would "catch up" to the download and be forced to wait. (To be sure, there was always *some* fear involved; see Murphy's Law.)

This breakthrough meant that most media would start playing within a few seconds. It also meant that, usually, the media was never really "downloaded" to the user's computer; instead, when the media started playing, it would continue to play until it was over or the user moved on to something else. Consuming media online became like sitting on the bank of a stream watching the water flow by, so some sensitive poet of an engineer coined the term *streaming* to describe this new way of listening to and viewing media.

Nowadays, streaming has gone, well, mainstream for a couple of reasons. First, many users now have computers and/or mobile devices that are powerful enough to process even the largest and most complex incoming audio and video signals. Second, lots of people (at least in the developed world) have speedy Internet connections and home networks, which means that streams usually start within a few seconds and the streaming *buffer* — the area of memory that's used to store the next few seconds or minutes of the media (more on this a bit later) — is always full, which results in a continuous and glitch-free playback.

Streaming today usually comes in one of the following forms:

>> **Audio streaming:** Mostly prerecorded music through services such as Amazon Music and Spotify, as well as podcasts through services such as myTuner Radio and Plex.

>> **Video streaming:** Mostly prerecorded TV shows and movies through services such as Amazon Prime Video and Netflix.

>> **Live streaming:** As-it's-happening audio or video, such as on-the-air TV programs delivered by your cable provider or Fire TV Recast, live concerts or sporting events, Internet-based audio or video phone calls, or video feeds of a specific place or scene.

Getting Clear on Streaming Media Devices

Imagine yourself sitting on the bank of a fast-running stream. Your eyes see the water whooshing by; your ears hear the babbling of the brook; your nose smells the wonderful scent of clean water; if you feel like it, you could also use your hands or feet to sense the coolness of the stream and your tongue to taste the freshness of pure water. In much the same way that your senses give you "access" to a stream, you need a special device to "access" an online media stream.

Sometimes that device is just a piece of software. For example, when you click to play a YouTube video, the YouTube site streams that video using special playback software that runs right in your web browser.

Increasingly these days, however, that device is a piece of hardware called a *streaming media device*, and it offers two main features:

>> **Streaming service interface:** A method for discovering and interacting with services that offer audio, video, or live streams. This feature is incredibly useful because there are dozens — nay, *hundreds* — of streaming services out there, so having a way to bring all your favorite services together in a single interface is mind-blowingly convenient.

>> **Streaming media playback:** The capability of playing, pausing, rewinding, and fast-forwarding an incoming media stream, usually by pressing buttons on a remote control that comes with the streaming media device.

Amazon gadgets such as the Fire TV Stick, Fire TV Stick 4K, and Fire TV Cube are streaming media devices that use your TV or a mobile device to display a streaming service interface and play audio, video, and live streams, which you can control using either the bundled remote or Alexa voice commands.

Understanding How Streaming Works

As you might imagine, streaming media is a hideously complex bit of business that requires extremely sophisticated hardware and software to make everything work as well as it does. The good news is that you don't need to know anything about that complexity, so you can shut off all those alarm bells going off in your head. Instead, this section provides you with a very basic overview of how streaming performs its magic.

The general process for streaming a prerecorded audio or video file is illustrated in Figure 1-1.

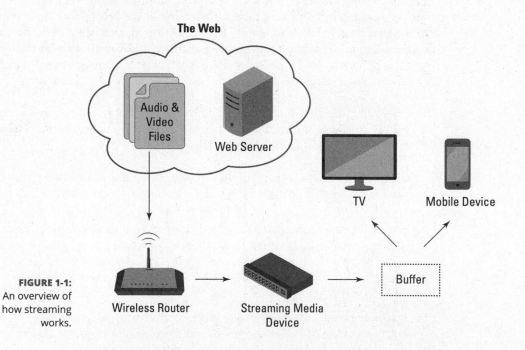

FIGURE 1-1:
An overview of how streaming works.

As Figure 1-1 shows, streaming is a five-step process

1. For prerecorded audio or video, the media file is stored on the web using a special computer called a *web server*.

2. When a user requests the media, the server begins sending the first few seconds of the audio or video file to the user.

3. When the data reaches the user's network, the network's wireless router passes the data along to the streaming media device.

   Note that the router is usually wireless, but it doesn't have to be.

4. The streaming media device waits until it has a certain amount of the media before it starts the playback.

   The saved data is stored in a special memory location called a *buffer*. (See the next section, "More about buffering," for, well, more about buffering.)

5. When the buffer contains enough data to ensure a smooth playback, the stream is sent to the user's TV or mobile device, and the entertainment begins.

More about buffering

The buffering process that occurs in steps 4 and 5 of the previous section is such a crucial part of streaming that it goes on throughout the playback, not just at the beginning. For example, when you examine the current progress of the playback, you usually see a progress bar that's similar to what I've illustrated in Figure 1-2. The circle shows your current position in the playback. Just ahead of the circle is a dark portion of the progress bar, which shows you how much of the upcoming stream is stored in the buffer; the rest of the progress bar is white, which tells you that part of the stream hasn't yet been received by the streaming media device. (The colors may vary on your TV or mobile device.)

Why not just play the media as it arrives and skip the buffer altogether? That would be nice, and it just might work in an ideal world, but the world we actually inhabit is far from ideal. In real life, media streams can suffer from a number of problems:

>> The server may be slow to respond if it has to deal with a large number of media requests.

>> Your Internet connection speed may be slow.

>> Your network speed may be slow.

>> Glitches between the server and your network may mean that large parts of the media stream are delayed or missing.

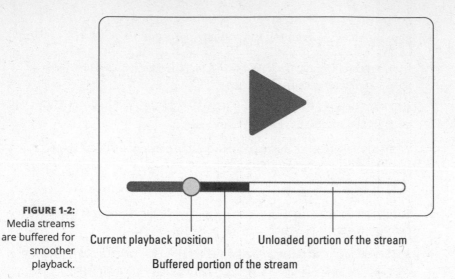

FIGURE 1-2:
Media streams
are buffered for
smoother
playback.

Current playback position　　　Unloaded portion of the stream

Buffered portion of the stream

Any one of these problems could cause the stream playback to be interrupted for anything from a split second to a few seconds. Without a buffer to fall back on, your show or song would have to stop mid-playback to wait for the delay to resolve itself. However, with anywhere from a few to a few dozen seconds stored in the buffer, the streaming media device can keep the stream playing, and you remain blissfully unaware of any problems because they happen in the background, without affecting your enjoyment of the media.

Streaming and data usage

When you're looking to sign up for an account with an Internet service provider (ISP), you're usually presented with several plans of varying prices. One of the features that varies with the price of each plan is the amount of data per month that you can transfer between the Internet and your modem. This is called *usage* or *monthly usage,* and the cheaper the plan, the lower the usage limit you get per month. That limit is important because if you go *over* that amount in a given month, the ISP will charge you a small fortune for each gigabyte (GB) that you exceed your cap.

WARNING

Overage fees can easily run to several dollars per gigabyte, and Internet forums are awash in tales of people getting dinged $100 or $200 for going way over their monthly allowance. So, yes, for all but the most well-heeled, this is big bucks I'm talking about here.

I'm telling you all this because it's important to know early in your streaming career that streaming media is very data-intensive, meaning it requires tons of usage. For example, here are the data usage values for some popular music streaming services:

Music Service	Data Usage per Hour
Amazon Music	0.11GB
Apple Music	0.11GB
Google Play Music	0.14GB
Pandora	0.57GB
Spotify	0.14GB

Similarly, here are the data usage values for various types of video stream qualities (most video streaming services give you the option of streaming video in two or more of these values):

Video Quality	Data Usage per Hour
Low	0.3GB
Standard Definition (SD)	0.7 GB
High Definition (HD)	1.0GB to 3GB
4K Ultra-High Definition (UHD)	7GB

Based on these usage values, you can see that it wouldn't be that hard to use anywhere from 5GB to 10GB of streaming media a day. That translates to 150GB to 300GB a month, which is bad news, indeed, if your ISP's monthly usage cap is 100GB!

TIP

My advice? Get an unlimited Internet plan if you can afford it. If that's too pricey for your budget, then keep an eagle eye on your daily usage (most ISPs offer a tool that lets you view your daily usage). If you see that things are getting out of hand, usage-wise, dial back the streaming for a few days or a week to make sure you don't go over your cap.

Knowing What You Need to Stream: Apps and Hardware

What you need to get into the world of streaming varies widely depending on a number of factors, including what streams you want to check out, your budget, your tolerance for complexity, and your desire for convenience.

At the simplest end of the streaming world, all you need is an Internet connection and a web browser. In this bare-bones scenario, you surf to a streaming site (such as Netflix, Spotify, or YouTube), sign into your account (if the site requires an account, as most do), and then use the site's interface to find and play the streams you want. Many streaming sites offer free accounts (supported by the ads you're forced to view), so you can get into streaming without forking over any extra cash.

TIP

Jumping around from one streaming website to another isn't hard, but it tends to get cumbersome as your stable of streaming services gets larger (as it inevitably will). One easy way to get around the inconvenience of multiple websites is to use the apps that most streaming services provide. Install the app on your smartphone or tablet, use the app to sign in to your account on the streaming service, and — *voilà!* — you can now locate, curate, and watch streaming media on your mobile device!

Viewing streaming media on a mobile device might seem like the perfect solution for an age obsessed with its smartphones and tablets. There's also something wonderfully intimate about watching a TV show or movie on a mobile device. That said, scratching the streaming itch using only mobile device apps does have its downsides:

» You may end up with apps scattered across multiple devices.

» If you're not connected to Wi-Fi, overindulging your mobile streaming media habit may put you over your cellular plan's monthly usage cap.

» You may prefer to watch a particular show on a large TV screen.

» It's hard for multiple people to watch a stream on a relatively small mobile screen.

TIP

There are ways to send a mobile device's video stream to a TV (this is called *mirroring*), but an easier way to solve all these problems is to throw more hardware at them. I speak, in particular, of a streaming media device, which, as I describe earlier, is designed to take an incoming stream and display it on a connected TV screen (or mobile device). Even better, all streaming media devices come with a large collection of apps for streaming services. For example, Figure 1-3 shows a portion of the Fire TV Home screen, which includes a row labeled *Your Apps & Games*. In the example, this row includes apps for the streaming services Amazon Prime Video, Netflix, TuneIn Radio, YouTube, and TED TV.

FIGURE 1-3:
Some typical
apps on the Fire
TV Home screen.

This means that you can collect all your streaming apps in one location and view each app's streaming content on whatever TVs and mobile devices you've connected to the streaming media device. It's a pretty sweet setup, and it's the Fire TV way of doing things. I dive into the Fire TV details starting in Chapter 2.

Chapter **2**

Getting to Know Fire TV

"**W**ell begun is half done" said some guy named Aristotle more than 2,000 years ago. He meant, I think, that any enterprise that gets off to a proper start stands a better chance of succeeding than an endeavor that has a rocky, uncertain, or downright erroneous beginning.

Far be it from me to argue with an ancient Greek philosopher, so my goal in this chapter is to get you "well begun" on your quest to not only learn the basics of Fire TV, but also get the best possible return on your Fire TV investment.

To that worthy end, in this chapter you explore the Fire TV landscape from hot-air-balloon height. From this perspective, you learn what Fire TV is, how you get it, and how to choose a Fire TV device if you haven't jumped on the bandwagon as yet. You also take a tour of what Fire TV can do and learn how the Alexa voice assistant fits into everything.

Getting Acquainted with Fire TV

People who are new to Fire TV usually find it very hard to grasp exactly what Fire TV is. So, if you're struggling to understand Fire TV, rest assured that you aren't struggling alone.

The best way I can think of to introduce you to Fire TV is to give you a bunch of analogous media scenarios:

>> When you connect a Blu-ray player to your TV, you can use the player to watch Blu-ray movies on your TV.

>> When you connect a CD player to your audio receiver, you can use the player to listen to music CDs through your sound system.

>> When you connect a digital camera to your computer, you can use the camera to view its photos on your computer's monitor.

The common thread in all these scenarios is that when you want to watch (or listen to or view) a particular type of media, you connect one device that can play the media to another device that can show the media.

Fire TV is another example of a device that can "play" media, and it uses your television to show that media. What type of media does Fire TV play, you ask? That's an excellent question, and I answer it in the next section.

Understanding What Fire TV Does

Watching television used to be simple: You turned on the set, tuned in to the channel you wanted to watch, and then settled in with your popcorn. Today, however, watching anything is complicated because content is scattered across multiple sources. One show is on Hulu, a second show is on Amazon Prime, another is on Netflix, and yet another resides on YouTube. Trying to keep it all straight is headache-inducing, and trying to coordinate your various devices, service addresses, and account logins is stressful.

To solve all these problems, Amazon created Fire TV. No, despite the name, Fire TV is not a television. Instead, it's a streaming media device (a term I define in Chapter 1) that connects to your existing TV. The purpose of Fire TV is to bring together most of the popular content services into a single interface on your TV. Connect Fire TV to your TV, switch to the Fire TV input, and you get quick access to apps for not only Amazon Prime, but also ESPN, HBO, Hulu, Netflix, Showtime, YouTube, and many more (more than 100 in all). You can also use Fire TV to play games, run apps, view photos, watch Internet videos, and more. Lots of devices and services claim to be a "one-stop shop" for content, but Amazon's Fire TV actually lives up to that billing.

Fire TV components

Fire TV isn't terribly complex, but for the whole streaming media thing to work, it does require a few components. Here's an overview of what you need to make the streaming magic happen:

» **Streaming media device:** To state the obvious, the most important component in any Fire TV system is Fire TV itself. Note, however, that there are two main flavors of Fire TV to consider:

- **Fire TV device:** This version of Fire TV is a separate device. As I write this in late 2019, Amazon offers three Fire TV devices: Fire TV Stick, Fire TV Stick 4K, and Fire TV Cube.

- **Fire TV Edition:** This version of Fire TV is built into another device, which is usually a TV set, but it can also be a soundbar (which is basically a sophisticated set of speakers rolled into a single, bar-shaped device).

» **TV set:** If you have a separate Fire TV device or a Fire TV Edition Soundbar, then you also need a TV to connect that device. Note that you can't use just any TV set with Fire TV — the only way to connect Fire TV is with an HDMI cable, so your TV must have an HDMI port (see Figure 2-1). All reasonably modern TVs come with at least one HDMI port (and the latest TVs have two or more), but if your TV is old and doesn't have an HDMI port, then no Fire TV for you.

FIGURE 2-1:
Fire TV devices require an HDMI port on your TV.

HDMI port

» **Internet connection:** Fire TV gets most of its content via the Internet. (I say "most" here because Fire TV Edition Smart TVs can also view live TV via a cable or antenna connection and you can also view photos and videos via an external storage drive.) Therefore, you can't do much with Fire TV without Internet access. To that end, you need your home network to be connected to the Internet, and you then connect Fire TV to your network via Wi-Fi (for kicks, a wired connection directly to your router is also possible).

>> **Apps:** It's important to understand that Fire TV itself doesn't come with any content. Instead, Fire TV gets all its content from streaming services such as Amazon Prime Video, Netflix, and YouTube. How does Fire TV get that content? Through its collection of apps, each of which is provided by a streaming service and is a small program designed to access, play, and control the service's streaming media. Many apps (such as Amazon Prime Video, if you're an Amazon Prime member, and YouTube) are free, but most others (such as Hulu and Netflix) require paid subscriptions.

How Fire TV works

Given the various Fire TV components that I outline in the preceding section, here's the general procedure that happens when you interact with Fire TV to play some media:

1. **Turn on your TV.**

2. **If you're using a separate Fire TV device, change your TV's input or source to the HDMI connection used by Fire TV.**

   You see the Fire TV interface.

   Note: If you're using a Fire TV Edition Smart TV, you see the Fire TV interface automatically when you turn on the set.

3. **Use the Fire TV interface to choose the app for the streaming service you want to view.**

   If this is the first time you've opened the app and the app requires an account, you're prompted to sign in to your account (or create an account, if you don't already have one).

4. **Sign in to your streaming service account, if required.**

   The app displays the interface for the streaming service.

5. **Use the app interface to choose the content you want to view.**

   The app plays the content, which appears on your TV screen.

6. **Use the buttons (such as Play/Pause, Rewind, Fast Forward, Volume Up, and Volume Down) on the Alexa Voice Remote to control the media playback.**

Figuring Out Which Fire TV Device You Need

To help you make the right Fire TV decision, in this section I offer a quick look at what's available from both Amazon and third-party manufacturers. Then I help you choose the right device for your needs.

First, the devices:

» **Fire TV Stick:** Fire TV Stick (shown in Figure 2-2) is the simplest Fire TV device because it's designed to work with TVs that offer High Definition (HD) picture-quality resolutions of 720p or 1080p. (No idea what "720p" and "1080p" mean? Check out the sidebar titled "TV resolutions: The big picture," a little later in this chapter to find out.)

Fire TV remote

Fire TV Stick

FIGURE 2-2: Amazon's Fire TV Stick.

Photograph courtesy of Amazon

» **Fire TV Stick 4K:** Fire TV Stick 4K (shown in Figure 2-3) is designed to work with today's higher-resolution TVs that offer up to 4K Ultra-High Definition (UHD) picture quality. (Again, see the sidebar titled "TV resolutions: the big picture" to find out what "4K UHD" means.)

FIGURE 2-3: Amazon's Fire TV Stick 4K.

Photograph courtesy of Amazon

» **Fire TV Cube:** Fire TV Cube (shown in Figure 2-4) is the most powerful Fire TV device. Not only does it support up to 4K UHD screen resolution, but it also comes with a built-in speaker, built-in voice recognition (meaning you can use the Alexa voice assistant to control your TV), and double the internal storage of the Fire TV Stick and Fire TV Stick 4K.

FIGURE 2-4:
Amazon's
Fire TV Cube.

Photograph courtesy of Amazon

» **Fire TV Edition Smart TV:** A Fire TV Edition Smart TV (see Figure 2-5) is probably the easiest way to get into the Fire TV game because it doesn't require anything other than the TV — the Fire TV interface is built in and appears automatically when you turn on the set. Nice.

» **Fire TV Edition Soundbar:** A Fire TV Edition Soundbar (see Figure 2-6) has Fire TV built in and provides that interface to your TV via an HDMI cable connection. The soundbar typically also comes with powerful speakers and a subwoofer for ridiculously impressive sound from such a compact package.

TIP

Okay, so which Fire TV device should you choose? That depends on your needs and budget, so here are some questions to help you decide:

» **Are you on a tight budget?** Go for the Fire TV Stick, which is the cheapest of the Fire TV devices.

» **Do you have a new (or newish) TV that supports 4K Ultra HD?** Go for the Fire TV Stick 4K, which is just a few dollars more than the Fire TV Stick, but you get awesome picture quality on shows that support 4K.

» **Do you want to use Alexa to control not only your TV, but also other smart-home devices?** Although you can do all that with a Fire TV Stick, you'd be better off springing for the Fire TV Cube, which responds to your voice commands even from across the room.

» **Are you looking to buy a new TV?** Consider a Fire TV Edition Smart TV, which comes with Fire TV baked in, so you get to enjoy Fire TV as soon as you turn on your set.

» **Do you want the best sound possible?** Go for a Fire TV Edition Soundbar, which offers rock-the-house sound without taking up half your living room.

FIGURE 2-5:
A Fire TV Edition Smart TV.

Photograph courtesy of Amazon

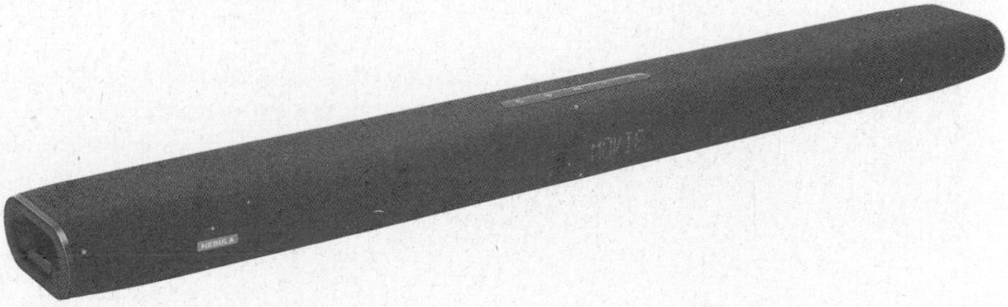

FIGURE 2-6:
A Fire TV Edition Soundbar.

Photograph courtesy of Amazon

TV RESOLUTIONS: THE BIG PICTURE

What's all this about the *resolution* of a TV screen? Briefly, the resolution determines how sharp the picture will appear. The keys here are the *pixels* (short for *picture elements*), which are the thousands of teeny pinpoints of light that make up the picture display. Each pixel shines with a combination of red, green, and blue, which is how they produce all the colors you see.

The important figures for TV resolution are the *horizontal resolution* and the number of *scan lines.* The horizontal resolution is the number of pixels across the screen. The scan lines are the horizontal lines created by these pixels. The number of scan lines is also called the *vertical resolution.* Basically, the higher these numbers are, the better the picture will be.

Old TV sets used a mode called NTSC (which, if you must know, stands for *National Television System Committee*) and had a horizontal resolution of 720 pixels with 486 scan lines. This is often written as 720 x 486, and multiplying these numbers together, it means the set had 349,920 total pixels to display each frame. Nowadays, even cheap TVs come with HD (sometimes called HDTV) resolution, which is 1920 x 1080, which multiplies out to 2,073,600 pixels, or about six times the NTSC value. That, in a nutshell, is why HD looks so much better than NTSC. A second HD format is 1280 x 720.

Other terms related to resolution that TV sales types bandy about are *interlaced scanning* (or just *interlacing*) and *progressive scanning.* Both refer to how the set "draws" each video frame on the screen. Inside the set is an electron gun that shoots a beam that runs along each scan line and lights up the pixels with the appropriate colors. With interlaced scanning, the beam first paints only the odd-numbered scan lines and then starts again from the top and does the even-numbered scan lines. With progressive scanning, the beam paints all the lines at once. In general, progressive scanning is better because it produces a more stable picture.

A resolution that uses progressive scanning over 1,080 scan lines is called *1080p,* while a resolution that uses progressive scanning over 720 scan lines is called *720p.* (For the record, the corresponding interlaced resolutions are known as *1080i* and *720i.*) If you see a set advertised as "HD (or HDTV) capable," it means it supports both 1080p and 720p.

Fire TV Stick 4K and Fire TV Cube also support Ultra HD, which is a resolution of 3840 x 2160. This means a horizontal resolution of 3,840 pixels (hence, this resolution is often called 4K UHD) and 2,160 scan lines. Ultra HD uses about four times the number of pixels as 1080p HD does.

Learning What Fire TV Can Do

Some people think a Fire TV device is nothing but a fancy-schmancy method for watching YouTube videos. Yep, sure, you can watch all the YouTube you can stand using Fire TV, but there's more — *much* more — that Fire TV can do. Let me show you what I mean.

Watching movies and TV shows

Because Fire TV connects to your television, it won't come as any surprise to learn that most people use Fire TV as a one-stop-shop for accessing movies and TV shows. Fire TV comes with apps from all the major streaming services, including free services such as Tubi and YouTube, and paid subscription services such as Hulu and Netflix. (See Chapter 6 to learn how to watch movies and TV shows with Fire TV.)

REMEMBER

Depending on where you live, your Fire TV device also comes with apps specific to your location, such as the Canadian Broadcasting Corporation (CBC) for Canadian users and the British Broadcasting Corporation (BBC) for users in the United Kingdom.

Accessing other types of media

Fire TV comes with apps that enable you to access a wide variety of other media, including the following:

>> **Music and other audio:** Fire TV has apps for most popular music services, including Amazon Music and Spotify, which gives you access to millions of songs right there on your TV. Fire TV can also tune in to radio stations and play podcasts.

>> **Photos:** Use the Amazon Photos app on your mobile device to load your photos to Amazon, and you can then view those photos using the Fire TV Amazon Photos app. Sweet.

>> **Games:** There are apps for hundreds of games that you can play right on your TV. Game categories include action, adventure, arcade, board, and cards.

>> **Internet:** You can access other media content online using web-browsing apps such as Amazon Silk and Firefox.

>> **Apps:** Fire TV boasts thousands of apps that you can access on your TV in categories ranging from education and finance to entertainment and lifestyle. There's something for everyone.

I talk more about accessing all these media types with Fire TV in Chapter 8.

Watching and recording over-the-air TV

If you add a Fire TV Recast device to your entertainment system, you can not only view over-the-air TV shows using a connected HDTV antenna, but also record those shows to watch later. It's a slick setup, and I tell you how it works in Chapter 7.

Connecting devices

Fire TV usually doesn't require anything extra, but if the need arises, you can augment your Fire TV system with either Bluetooth or USB devices. For example, if you want to watch TV without disturbing people nearby, you can connect a pair of Bluetooth headphones to your TV; if you want to view your own videos and photos on Fire TV, you can connect a USB storage drive. I go into just the right amount of detail about all this in Chapter 8.

Controlling your TV with Alexa voice commands

Alexa is Amazon's personal voice assistant, which means you can get information and make things happen (such as playing music) with just voice commands. Alexa is built right into the Fire TV experience, which means you can control your TV, navigate the Fire TV interface, and control the playback of movies, TV shows, and other content with just a few, simple voice commands. It's cool in the extreme, and I take you through the details in Chapter 9.

Controlling your smart home

Alexa is smart-home savvy, so it gives you voice control over many different home automation products, including lights, thermostats, baby monitors, security cameras, and door locks. I tell you how to set up and manage your smart home in Chapter 9.

IN THIS CHAPTER

» **Starting with the input source**

» **Getting a Fire TV Stick or Fire TV Stick 4K up and running**

» **Adding a Fire TV Cube to your home entertainment system**

» **Getting Fire TV Edition on the job**

Chapter **3**

Setting Up Fire TV

If you're a certain age, you may remember when devices were advertised as being "plug-and-play," which meant, at least in theory, that all you had to do was connect the device and it would configure itself automatically, meaning you could then "play" with the device (whatever that meant) after a minute or two. (Note I said "in theory"; in practice, such devices were better described as "plug-and-pray.")

I'm sorry to report that your Fire TV device does not fall under the "plug-and-play" rubric. Instead, after you plug in your Fire TV device, there's a non-trivial configuration process that you must run through before you can play with it. That process includes crucial steps such as connecting to your Wi-Fi network and signing in to your Amazon account. Lucky for you, the entire process takes only a few minutes, and this chapter takes you through every step.

Changing Your TV's Input Source

As you learn in this chapter, if you have a Fire TV Stick, Fire TV Stick 4K, or Fire TV Cube, you must connect the device to an HDMI port on your TV. Here's a question that stumps many people who are new to Fire TV: How does your TV "know" that Fire TV is connected to its HDMI port? That is, when you turn on your TV, you just see your regular TV channels — how do you see the Fire TV stuff, instead?

All modern TVs have a feature called the *input source,* which refers to the incoming connection that the TV uses to display a signal on its screen. By default, your TV uses the input source for your cable (or satellite or antenna) connection, so you see your regular channels. To see a different incoming signal — which might be for a connected device such as a Blu-ray player — you need to choose the appropriate input source. More to the point for this chapter, after you've connected your Fire TV device to an HDMI port on your TV (and to a power outlet), your next task is to turn on your TV and change the input source to the Fire TV device's HDMI port.

How you change the input source depends on the TV, but in most cases you use the TV's remote to press a button labelled Input (or, in some cases, Source), as shown in Figure 3-1. You then change the input source in one of two ways:

>> Keep pressing the button until you've selected the input source you want to view.

>> A menu of available input sources appears and you use the remote's navigation ring or buttons to select the source you want.

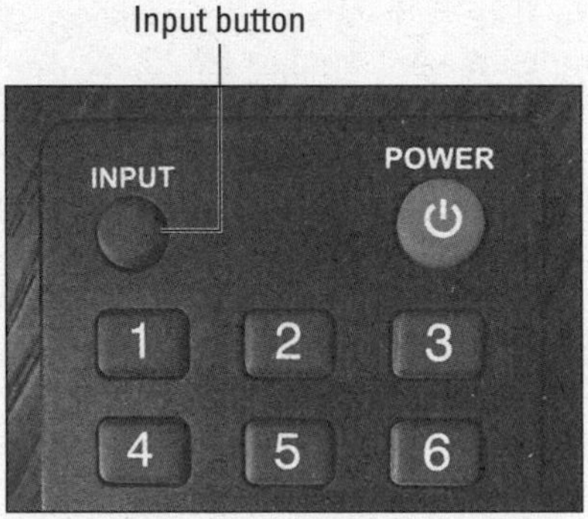

Input button

FIGURE 3-1:
On most TV remotes, you change the input source by pressing the Input (or Source) button.

TIP   If you don't see a button named Input (or Source) on your remote, or if you don't have a remote, look on the TV itself, which should have an Input (or Source) button.

Setting Up Fire TV Stick or Fire TV Stick 4K

If you have a Fire TV Stick or Fire TV Stick 4K, you need to connect the device to your TV and then run through the setup procedure. The next two sections take you through all the steps.

In the rest of this section, instead of saying the long-winded phrase *Fire TV Stick or Fire TV Stick 4K*, I'm going to save wear-and-tear on my typing fingers (and your reading eyes) by shortening that phrase to *Fire TV Stick.*

Connecting Fire TV Stick to your TV

Your Fire TV Stick connects to your TV's HDMI port, which on most TVs is labeled *HDMI.* If your TV has multiple HDMI ports (as most modern TVs do), then the ports are usually labeled HDMI 1, HDMI 2, and so on, as shown in Figure 3-2. Newer TVs usually have all their HDMI ports on one side of the TV's back panel (refer to Figure 3-2), while on older TVs it's common to have one HDMI port on the bottom of the TV's back panel and a second HDMI port on the side of the back panel (see Figure 3-3).

HDMI ports

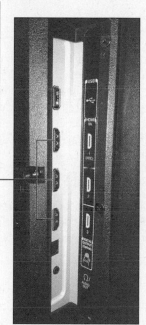

FIGURE 3-2: Modern TVs have all their HDMI ports together on the back panel.

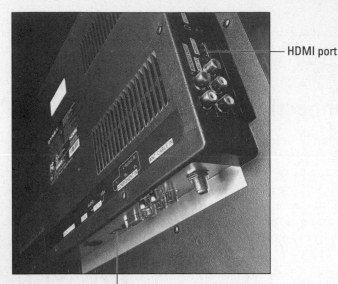

HDMI port

HDMI port

FIGURE 3-3:
Older TVs often
have their HDMI
ports in multiple
locations on the
back panel.

The location of the HDMI port is important because the length of the Fire TV Stick (especially the longer Fire TV Stick 4K) often means that there isn't room between a bottom HDMI port and whatever surface the TV is sitting on for the Fire TV Stick to fit. If that's the case for you, then you have three possible solutions:

>> Plug the Fire TV Stick into a side HDMI port, if you have one available.

>> Mount the TV on the wall (which gives the Fire TV Stick plenty of room because there's no longer a surface immediately under the TV).

>> Use the HDMI extender cable (see Figure 3-4) that came with your Fire TV Stick. In this case, insert the smaller end of the extender cable into the HDMI port on your TV; then connect your Fire TV Stick to the larger end of the extender cable (refer to Figure 3-4).

With your Fire TV Stick connected to your TV, grab the USB cable that came with your Fire TV Stick. Connect one end of the USB cable to the port on the side of the Fire TV Stick, plug the other end of the USB cable into the USB port on the power adapter that came with your Fire TV Stick, and then plug the power adapter into a power outlet.

Turn on your TV and change the input source (as I describe earlier in the "Changing Your TV's Input Source" section) to the Fire TV Stick HDMI connection.

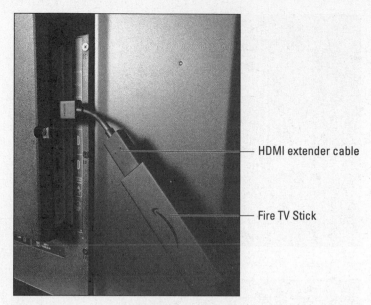

HDMI extender cable

Fire TV Stick

FIGURE 3-4:
A Fire TV Stick
with the HDMI
extender cable
attached.

Setting up Fire TV Stick

Your Fire TV Stick comes with an Alexa Voice Remote that you use to navigate the Fire TV interface (either by pressing buttons or by using voice commands). Before proceeding with the Fire TV Stick configuration, remove the back cover of the Alexa Voice Remote, insert the two batteries that came with Fire TV Stick, and then reattach the back cover.

When you tune to the Fire TV Stick HDMI port on your TV, the Fire TV Stick starts up for the first time and takes you through the following setup process:

1. **When you see the Searching for Your Remote message, as shown in Figure 3-5, press the Home button on the Alexa Voice Remote.**

Pressing the Home button enables the Fire TV Stick and the Alexa Voice Remote to connect (or *pair*) with each other.

TIP

If nothing happens when you press the Home button, you can force the Alexa Voice Remote into pairing mode by pressing and holding the Home button for ten seconds, or until you see the Alexa Voice Remote LED rapidly flashing amber.

When the pairing is complete, you see the message shown in Figure 3-6.

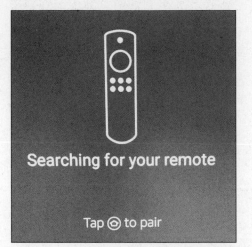

FIGURE 3-5:
When you see
this message,
press the Home
button on your
Alexa Voice
Remote.

Searching for your remote

Tap ⊙ to pair

fire tv stick

FIGURE 3-6:
You see this
message when
the pairing
between the Fire
TV Stick and the
Alexa Voice
Remote is
complete.

Press ⊙ to start

2. **On the Alexa Voice Remote, press the Play/Pause button.**

   Fire TV Stick asks you to choose your language.

3. **Use the Alexa Voice Remote to choose the language you want to use.**

   I talk about how to use the Alexa Voice Remote in more detail in Chapter 4. For now, you use the Alexa Voice Remote to "choose" something by using the navigation ring to press Down (the bottom part of the ring), Up (the top part of the ring), Left (the left part of the ring), or Right (the right part of the ring) to highlight the item you want; then press Select (the circular area in the middle of the navigation ring).

   After a few moments, Fire TV Stick displays a list of nearby Wi-Fi networks.

4. **Choose your Wi-Fi network.**

   Fire TV Stick prompts you to enter the network password.

5. **Use the Alexa Voice Remote navigation ring to enter each character in your network password, and then choose Connect (or press Play/Pause on the remote).**

TIP

   If your network password requires one or more uppercase letters, you can switch to uppercase characters by choosing the aA button or by pressing Menu on the Alexa Voice Remote.

   Fire TV Stick connects to your network and then checks to see if there is an available update to Fire OS, the operating system that runs Fire TV Stick. If an update is available, Fire TV Stick downloads and installs the software, which can take a few minutes. Note that Fire TV Stick may restart during this process.

   Eventually, Fire TV Stick asks you to sign in to your Amazon account.

6. **Choose I Already Have an Amazon Account.**

   Fire TV Stick displays the Enter Your Amazon Login ID screen.

   What if you don't have an Amazon account? No problem. Choose I Am New to Amazon; then use the Create Account screen to set up your account.

7. **In the Email Address field, use the Alexa Voice Remote navigation ring to type your Amazon account's email address, and then choose Next.**

   Fire TV Stick displays the Enter Your Amazon Account Password screen.

8. **In the Password field, use the Alexa Voice Remote navigation ring to type your Amazon account's password (see Figure 3-7), and then choose Sign In.**

   By default, Fire TV Stick hides the password by displaying each character as a dot, as shown in Figure 3-7. If you want to make sure you entered the password correctly, choose the Show Password button.

REMEMBER

   If you've enabled two-factor authentication on your Amazon account (as I describe in Chapter 12), Fire TV Stick will prompt you to enter a code to verify the sign-in. Type the code that was sent to you, and then choose the Next button.

   Fire TV Stick confirms your Amazon credentials, signs in to your account, and then registers your Fire TV Stick. Fire TV Stick then asks which Amazon account you want to use.

Enter Your Amazon Account Password   2 | 2

1	2	3	4	5	6	7	8	9	0
a	b	c	d	e	f	g	h	i	j
k	l	m	n	o	p	q	r	s	t
u	v	w	x	y	z	!	,	.	@

⊜ aA   #$%   äçé   ⊙ Space   ⊙ Delete   **Show Password**

⤺ Previous   ⊙ SIGN IN

FIGURE 3-7:
Enter your
Amazon account
password.

9. **Choose Continue.**

Fire TV Stick asks if you want to save your Wi-Fi password to Amazon. This is part of a feature that Amazon calls Wi-Fi Simple Setup, which enables other Amazon devices you own (such as Echo smart speakers) to automatically connect to your network. This feature really makes setting up those devices easier, so it's a good idea to let Amazon save your password.

10. **Choose Yes.**

Fire TV Stick now prompts you to enable parental controls. I talk about parental controls in detail in Chapter 8, so you can skip this part of the setup for now.

11. **Choose No Parental Controls.**

Fire TV Stick next makes sure the volume buttons on the Alexa Voice Remote are working properly. Before continuing, make sure your TV's volume is turned up.

12. **Choose Next.**

Fire TV Stick plays some music so that you can test the Alexa Voice Remote volume buttons.

13. **On the Alexa Voice Remote, press the Volume Up (+) and Volume Down (–) buttons.**

Fire TV Stick asks if the music volume changed when you pressed the Alexa Voice Remote volume buttons.

If the volume didn't change, double-check that the TV volume is turned up loud enough that you can hear the music. Also, make sure you point the Alexa Voice Remote at your TV and that the TV's output device (if any) is turned on and connected properly.

14. **Choose Yes.**

Fire TV Stick tells you that the Alexa Voice Remote is configured.

15. **Choose OK.**

Fire TV Stick asks if you want to set up your streaming services.

I cover streaming in Chapter 6, so you don't need to set up streaming services right away.

16. **Choose No Thanks.**

If you do want to set up your services now, choose Get Started, and then follow the onscreen prompts.

Congratulations! Your Fire TV Stick is ready for action.

Setting Up Fire TV Cube

If you have a Fire TV Cube, you need to position the device optimally, connect the device to your TV, and then run through the setup procedure. The next few sections explain all.

Positioning your Fire TV Cube

After you've liberated your Fire TV Cube from its packaging, one obvious question arises: Where the heck do you put it? Somewhere near your TV seems like the obvious answer, but choosing the best location is a bit more complicated than that. Here are some things to consider:

» Your Fire TV Cube requires full-time power, so make sure there's an outlet close enough to the device.

» Your Fire TV Cube connects to an HDMI port on your TV, so it needs to be close enough to your TV that your HDMI cable can reach.

» TV or sound system speakers can befuddle the Fire TV Cube built-in microphone, so make sure all speakers are at least 1 to 2 feet away from your Fire TV Cube.

» Make sure the Fire TV Cube is within range of your Wi-Fi network.

» Make sure the device is close enough that you can give your voice commands without having to yell. Depending on the ambient noise in your environment, this usually means being within 15 to 20 feet of the device.

» Don't store the Fire TV Cube inside a cabinet or other enclosed location.

» Position the Fire TV Cube with the four buttons (see Figure 3-8) facing up, and the Amazon logo facing where you usually sit when you watch TV.

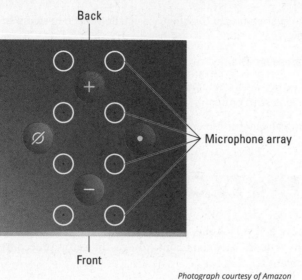

Back

Microphone array

FIGURE 3-8: Your Fire TV Cube uses an array of microphones to hear what you're saying.

Front

Photograph courtesy of Amazon

Getting to know your Fire TV Cube device's Alexa hardware

Fire TV Cube is a combination of a streaming media device and an Alexa-enabled smart speaker. The latter means that you can send voice commands to the Fire TV Cube to control your TV and Fire TV, and the Fire TV Cube also responds to questions and other utterances. I talk about Alexa in great detail in Chapter 9, but for now, before continuing with the Fire TV portion of the setup, it makes sense to first learn the basics of your Fire TV Cube device's Alexa hardware.

Taking a closeup look at the far-field microphone

Computers have had either external or built-in microphones for a few decades, and, of course, smartphones and tablets have had internal microphones from day one. But the characteristic that all these microphones have in common is that they assume the speaker is relatively close — within a few inches or, at most, a foot or two. Move much farther away, and those microphones get notoriously unreliable because they have trouble distinguishing your voice from the background noise in your environment.

That sort of second-rate microphone performance just won't do for the Fire TV Cube whenever you're relying on voice commands to get things done because those commands could be coming from 10, 15, or even 20 feet away. To get accurate and clear voice recordings, your Fire TV Cube relies on a technology called the *far-field microphone*, which is optimized to distinguish a voice from the ambient room noise even when that voice is far away. The Fire TV Cube far-field microphone uses some fancy-schmancy technology to accomplish this difficult task:

>> **Microphone array:** The Fire TV Cube "microphone" is actually an array of eight individual microphones, as shown by the circles in Figure 3-8. Note that these microphones are arranged somewhat narrowly from the front of the device to the back. The line created by this arrangement is the direction that Fire TV Cube expects your voice commands to come from. This is why, as I explain earlier (see "Positioning your Fire TV Cube"), you need to place your Fire TV Cube so that the front (where the Amazon logo resides) is pointing to where you sit when you watch TV.

>> **Noise reduction:** Detects unwanted audio signals (known in the audio trade as *noise*) and reduces or eliminates them.

>> **Acoustic echo cancellation:** Detects sounds coming from a nearby loudspeaker (such sounds are known as *acoustic echo*) — even if that loudspeaker is the Fire TV Cube itself — and reduces or cancels them to ensure accurate voice recordings.

>> **Beamforming:** Uses the microphone array to determine the direction your voice is coming from and then uses that directional information to home in on your voice.

>> **Barge-in:** The microphone ignores whatever media the Alexa device is currently playing — such as a song or podcast — so that the microphone can more easily detect and recognize a simultaneous voice command (thus enabling that command to "barge in" on the playing media).

>> **Speech recognition:** Detects the audio patterns associated with speech and focuses on those patterns instead of any surrounding noises.

Pushing the Fire TV Cube buttons

Your Fire TV Cube is built to be a hands-free device, which is a welcome design choice when you have one hand in a bowl of popcorn and the other clutching your favorite beverage. However, *hands-free* doesn't mean *hands-off* because the outer shell of your Fire TV Cube is festooned with four buttons that you can use to control certain aspects of the device.

Figure 3-9 shows the top of a Fire TV Cube and points out the four buttons.

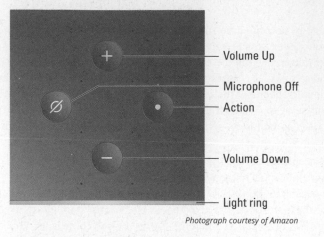

Volume Up

Microphone Off

Action

Volume Down

FIGURE 3-9:
The buttons that
dot the top of the
Fire TV Cube.

Light ring

Photograph courtesy of Amazon

Here's a summary of the available buttons:

» **Action:** Activates Alexa. That is, pressing the Action button is the same as saying Alexa's wake word.

» **Microphone Off:** Turns the Fire TV Cube microphone off. When the microphone is off, this button glows red, as does the Fire TV Cube light ring. Press this button again to turn the microphone back on.

» **Volume Up:** Increases the volume (no surprise, there).

» **Volume Down:** You guessed it: decreases the volume.

Connecting Fire TV Cube to your TV

Your Fire TV Cube connects to your TV's HDMI port (see Figure 3-10), which on most TVs is labeled HDMI. If your TV has multiple HDMI ports (as most modern TVs do), then the ports are usually labeled HDMI 1, HDMI 2, and so on (refer to Figure 3-2).

Your Fire TV Cube has a built-in infrared transmitter that enables you to use voice commands to control other devices in your home entertainment system, such as your TV (for example, to turn it on and off), audio receiver, Blu-ray player, cable set-top box, or satellite receiver. In other words, your Fire TV Cube, besides being a streaming media device and a smart speaker, is also a universal remote!

FIGURE 3-10:
Use an HDMI
cable to connect
your Fire TV Cube
to your TV.

TIP

The IR transmitter (some folks call it an *IR blaster*) is inside Fire TV Cube, but it may not work for some devices that you've stored inside a cabinet or other enclosed area. Are you out of the IR blaster game in that case? Not at all. Instead, you can connect the IR extender cable that comes with Fire TV Cube. Connect the IR extender cable's jack to the corresponding port on the back of Fire TV Cube, and then position the cube end of the cable as close as you can to the enclosed device or devices you want to control.

With your Fire TV Cube connected to your TV, grab the power cable that came with your Fire TV Cube. Connect one end of the power cable to the power port on the back of the Fire TV Cube, and plug the other end of the cable into a power outlet.

Turn on your TV and change the input source (as I describe earlier in the "Changing Your TV's Input Source" section) to the Fire TV Cube HDMI connection.

Setting up Fire TV Cube

Your Fire TV Cube comes with an Alexa Voice Remote that you can use to navigate the Fire TV interface. You may decide to opt only to use Alexa voice commands to control Fire TV, but you still need the remote to get through the initial configuration. So, before proceeding, remove the back cover of the Alexa Voice Remote, insert the two batteries that came with Fire TV Cube, and then reattach the back cover.

When you tune to the Fire TV Cube HDMI port on your TV, the Fire TV Cube starts up for the first time and takes you through the following setup process:

1. **On the Alexa Voice Remote, press the Play/Pause button.**

Fire TV Cube asks you to choose your language.

2. **Use the Alexa Voice Remote to choose the language you want to use.**

I talk about how to use the Alexa Voice Remote in more detail in Chapter 4. For now, you use the Alexa Voice Remote to "choose" something by using the navigation ring to press Down (the bottom part of the ring), Up (the top part of the ring), Left (the left part of the ring), or Right (the right part of the ring) to highlight the item you want; then press Select (the circular area in the middle of the navigation ring).

After a few moments, Fire TV Cube displays a list of nearby Wi-Fi networks.

3. **Choose your Wi-Fi network.**

Fire TV Cube prompts you to enter the network password.

4. **Use the Alexa Voice Remote navigation ring to enter each character in your network password, and then choose Connect (or press Play/Pause on the remote).**

TIP

If your network password requires one or more uppercase letters, you can switch to uppercase characters by choosing the aA button or by pressing Menu on the Alexa Voice Remote.

Fire TV Cube connects to your network and then checks to see if there is an available update to Fire OS, the operating system that runs Fire TV Cube. If an update is available, Fire TV Cube downloads and installs the software, which can take a few minutes. Note that Fire TV Cube may restart during this process.

Eventually, Fire TV Cube asks you to sign in to your Amazon account.

5. **Choose I Already Have an Amazon Account.**

Fire TV Cube displays the Enter Your Amazon Login ID screen.

What if you don't have an Amazon account? No problem. Choose I Am New to Amazon; then use the Create Account screen to set up your account.

6. **In the Email Address field, use the Alexa Voice Remote navigation ring to type your Amazon account's email address, and then choose Next.**

Fire TV Cube displays the Enter Your Amazon Account Password screen.

7. **In the Password field, use the Alexa Voice Remote navigation ring to type your Amazon account's password, and then choose Sign In.**

By default, Fire TV Cube hides the password by displaying each character as a dot. If you want to make sure you entered the password correctly, choose the Show Password button.

REMEMBER

If you've enabled two-factor authentication on your Amazon account (as I describe in Chapter 12), Fire TV Cube will prompt you to enter a code to verify the sign-in. Type the code that was sent to you, and then choose the Next button.

Fire TV Cube confirms your Amazon credentials, signs in to your account, and then registers your Fire TV Cube. Fire TV Cube then asks which Amazon account you want to use.

8. **Choose Continue.**

Fire TV Cube now asks if you want to save your Wi-Fi password to Amazon. This is part of a feature that Amazon calls Wi-Fi Simple Setup, which enables other Amazon devices you own (such as Echo smart speakers) to automatically connect to your network. This feature really makes setting up those devices easier, so it's a good idea to let Amazon save your password.

9. **Choose Yes.**

Fire TV Cube now prompts you to enable parental controls. I talk about parental controls in detail in Chapter 8, so you can skip this part of the setup for now.

10. **Choose No Parental Controls.**

Fire TV Cube asks if you want to set up your streaming services.

I cover streaming in Chapter 6, so you don't need to set up streaming services right away.

11. **Choose No Thanks.**

If you do want to set up your services now, choose Get Started, and then follow the onscreen prompts.

Fire TV Cube asks if you want Alexa to help you set up your TV and other equipment.

12. **Choose Continue.**

If you don't want to bother with this, choose Do This Later, instead.

REMEMBER

To actually do this later, choose Settings ➪ Equipment Control ➪ Set Up Equipment.

13. Gather the remote controls for each device you want Fire TV Cube to control, and then choose Next.

You need the remotes both to confirm when Fire TV Cube has successfully controlled some aspect of the device (by pressing the Fast Forward button) and to perform certain tasks that Fire TV Cube wants to learn how to perform itself (such as changing the input source on your TV).

Fire TV Cube runs through your devices, trying each time to detect the device automatically. If Fire TV Cube detects an incorrect device, be sure to choose No and then select the correct device from the list that appears.

14. When the equipment setup is complete, choose Continue.

That's it! Your Fire TV Cube is ready to roll.

Setting Up Fire TV Edition

Your Fire TV Edition comes with an Alexa Voice Remote that you use to navigate the Fire TV interface (either by pressing buttons or by using voice commands). Before proceeding with the Fire TV Edition configuration, remove the back cover of the Alexa Voice Remote, insert the two batteries that came with Fire TV Edition, and then reattach the back cover.

If your Fire TV Edition comes via a device other than a Smart TV (such as a sound-bar), connect the device to your existing TV's HDMI port, as follows:

» If your device and your TV support Audio Return Control (ARC), connect the device to your TV's HDMI (ARC) port. Otherwise, just connect the device to a regular HDMI port. Either way, be sure to select the Fire TV source mode on the soundbar (on the Nebula soundbar, for example, this is the FTV source mode). Also, be sure to turn off your TV's built-in speakers (see your TV manual for instructions).

» If you use other inputs on your TV, connect the device to an audio port on the TV, such as an optical or AUX port. In this case, when you switch to the other input, be sure to select the optical or AUX source mode on the soundbar (on the Nebula soundbar, for example, you'd select the OPTIC or AUX source mode).

When you power up the Fire TV Edition Smart TV for the first time, or when you switch to the Fire TV Edition input if you're using a device such as a soundbar, Fire TV Edition takes you through the following setup procedure:

1. **Use the Alexa Voice Remote to choose the language you want to use.**

   I talk about how to use the Alexa Voice Remote in more detail in Chapter 4. For now, you use the Alexa Voice Remote to "choose" something by using the navigation ring to press Down (the bottom part of the ring), Up (the top part of the ring), Left (the left part of the ring), or Right (the right part of the ring) to highlight the item you want; then press Select (the circular area in the middle of the navigation ring).

   TIP

   The TV should automatically pair with the Alexa Voice Remote. If nothing happens when you press the Alexa Voice Remote buttons, you can force the Alexa Voice Remote into pairing mode by pressing and holding the Home button for ten seconds, or until you see the Alexa Voice Remote LED rapidly flashing amber.

2. **Choose Continue.**

   After a short delay, Fire TV Edition displays a list of available Wi-Fi networks.

3. **Choose your Wi-Fi network.**

   Fire TV Edition prompts you to enter the network password.

4. **Use the Alexa Voice Remote navigation ring to enter each character in your network password, and then choose Connect.**

   TIP

   If your network password requires one or more uppercase letters, you can switch to uppercase characters by choosing the aA button or by pressing Menu on the Alexa Voice Remote.

   Fire TV Edition connects to your network and then checks to see if there is an available update to Fire OS, the operating system that runs Fire TV Edition. If an update is available, Fire TV Edition downloads and installs the software, which can take a few minutes. Note that Fire TV Edition may restart during this process.

   Eventually, Fire TV Edition asks which Fire TV experience you want to have.

5. **Choose Full to get the most out of Fire TV.**

   Fire TV Edition asks you to sign in to your Amazon account.

6. **Choose I Already Have an Amazon Account.**

   Fire TV Edition displays the Enter Your Amazon Login ID screen.

   What if you don't have an Amazon account? No problem. Choose I Am New to Amazon, and then use the Create Account screen to set up your account.

7. **In the Email Address field, use the Alexa Voice Remote navigation ring to type your Amazon account's email address, and then choose Next.**

   Fire TV Edition displays the Enter Your Amazon Account Password screen.

8. **In the Password field, use the Alexa Voice Remote navigation ring to type your Amazon account's password, and then choose Sign In.**

By default, Fire TV Edition hides the password by displaying each character as a dot. If you want to make sure you entered the password correctly, choose the Show Password button.

REMEMBER

If you've enabled two-factor authentication on your Amazon account (as I describe in Chapter 12), Fire TV Edition will prompt you to enter a code to verify the sign-in. Type the code that was sent to you, and then choose the Next button.

Fire TV Edition confirms your Amazon credentials, signs in to your account, and then registers your Fire TV Edition device.

2

Watching Fire TV

to Know the Alexa Voice Remote

One of the most important doodads that comes with your Fire TV device is the Alexa Voice Remote, which enables you to control Fire TV in two ways:

>> **Manually:** Press the remote's buttons to navigate the Fire TV interface and control playback.

>> **By voice:** Use the remote's built-in Alexa voice assistant to issue verbal instructions for navigating the interface and controlling content.

Touring the Fire TV Alexa Voice Remote

The Alexa Voice Remote will be your constant companion while you use Fire TV, so it pays to take a couple of minutes now to learn the lay of the remote land.

First, Figure 4-1 shows the version of the Alexa Voice Remote that ships with devices such as Fire TV Stick, Fire TV Stick 4K, and Fire TV Cube.

Here's a summary of what you see on the face of the Alexa Voice Remote:

>> **Power:** Press this button to turn your TV on and off.

>> **Microphone:** The internal device that picks up your voice requests when you press and hold the Voice button.

>> **LED:** Lights up when the Alexa Voice Remote is performing certain operations, such as signaling the remote's readiness to be paired with a Fire TV device.

>> **Voice:** Press and hold this button to make a voice request, as I describe a bit later in this chapter (see "Introducing Voice Control of Fire TV").

>> **Navigation ring:** Use this ring to navigate the Fire TV interface (see "Navigating with the Alexa Voice Remote," later in this chapter).

>> **Select:** Press this button to choose the item that's currently highlighted on the Fire TV screen.

>> **Back:** Press this button to go back to the previous Fire TV screen.

>> **Home:** Press this button to jump directly to the Fire TV Home screen.

>> **Menu:** Press this button to see a menu of commands that are specific to the item that's currently highlighted on the Fire TV screen.

Chapter 4

Learning Fire T

A t this point in your Fire TV career
attached to your TV (unless you ha
no such attachment is necessary),
registered with your Amazon account. If yo
Fire TV to-do list, then I'm happy to rep
Fire TV!

In this chapter, you take the next step by lea
Fire TV, but also for Amazon's Fire TV mobil
Voice Remote, install the Fire TV mobile app,
As a bonus, you also learn a few basic voice
using Amazon's Alexa voice assistant. Crack y
get ready to start playing with Fire TV!

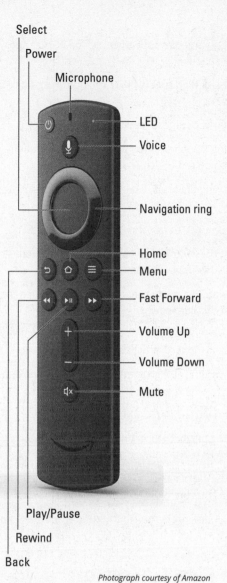

Select

Power

Microphone

LED

Voice

Navigation ring

Home

Menu

Fast Forward

Volume Up

Volume Down

Mute

Play/Pause

Rewind

Back

FIGURE 4-1:
The Alexa Voice
Remote that
ships with Fire TV
devices.

Photograph courtesy of Amazon

>> **Rewind:** During playback, use this button to rewind the content as follows:

- Press the button once to rewind by ten seconds.

- Press and hold the button for a few seconds, and then release the button to continuously rewind at low speed.

- While rewinding at low speed, press the button again to continuously rewind at medium speed.

- While rewinding at medium speed, press the button again to continuously rewind at high speed.

- While rewinding at high speed, press the button again to switch to the low-speed rewind.

- To stop rewinding, press the Play/Pause button.

» **Play/Pause:** During playback, press this button to pause the content; press the button again to restart the content.

» **Fast Forward:** During playback, use this button to fast-forward the content as follows:

- Press the button once to fast-forward by ten seconds.

- Press and hold the button for a few seconds, and then release the button to continuously fast-forward at low speed.

- While fast-forwarding at low speed, press the button again to continuously fast-forward at medium speed.

- While fast-forwarding at medium speed, press the button again to continuously fast-forward at high speed.

- While fast-forwarding at high speed, press the button again to switch to the low-speed fast-forward.

- To stop fast-forwarding, press the Play/Pause button.

» **Volume Up:** During playback, press this button to raise the volume; press and hold this button to raise the volume quickly.

» **Volume Down:** During playback, press this button to decrease the volume; press and hold this button to decrease the volume quickly.

» **Mute:** During playback, press this button to toggle the volume off and on.

Touring the Fire TV Edition Alexa Voice Remote

If you're doing the Fire TV thing using a Fire TV Edition Smart TV or Fire TV Edition Soundbar, then you get a slightly different configuration for the Alexa Voice Remote, as shown in Figure 4-2. This configuration has the same buttons as the Fire TV Alexa Voice Remote shown in Figure 4-1, but it also comes with the following extra buttons:

» **Guide:** Press this button to display the live TV channel guide (assuming you have live TV signals or apps; see Chapter 5).

» **Hot buttons:** These four buttons take you directly to the app for whatever streaming service is named on the button. For example, pressing the Netflix button opens the Netflix app. As a bonus, pressing one of the hot buttons also turns on your TV if it's currently off.

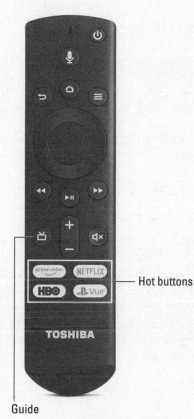

Hot buttons

FIGURE 4-2:
An example of an Alexa Voice Remote that ships with Fire TV Edition devices.

Guide

Photograph courtesy of Amazon

TIP

If you're accessing Fire TV Edition in a country other than the United States, your Alexa Voice Remote will likely offer one or more hot buttons that are different from the ones shown in Figure 4-2.

Checking Out the Fire TV Mobile App Remote

Although you'll mostly use the Alexa Voice Remote to control your Fire TV, if you lose the remote or it's not handy, you have an alternative: the Fire TV mobile app's Remote feature.

Installing the Fire TV mobile app

The Fire TV mobile app is a program that you download to your smartphone or tablet. You mostly use it to control a Fire TV device, but if you have Fire TV Recast as part of your system, you can also use the app to watch over-the-air TV, record shows, and watch those recordings (see Chapter 7 to learn how all this works).

Okay, so what do you need to get the Fire TV mobile app? A smartphone or tablet that meets one of these qualifications:

>> An iPhone or iPad running iOS 10 or later

>> An Android phone or tablet running Android 5 or later

>> An Amazon Fire tablet running Fire OS 3 or later

If you have one of these devices, go to your device's app store, search for the Amazon Fire TV mobile app, and install it.

When you first open the Fire TV mobile app, you're prompted to sign in with your Amazon account credentials. If you don't see that prompt, you can also tap the Sign In button at the bottom of the screen.

Figure 4-3 shows the Sign-In page for the Fire TV mobile app on an iPhone.

Follow these steps to sign in:

1. **In the Email field, type your Amazon account's email address.**

   REMEMBER

   If you use an Amazon mobile phone account instead of a standard Amazon account, type your mobile phone number into the field instead of your email address.

   What if you don't have an Amazon account? No problem. Tap the Create a New Amazon Account button; use the Create Account screen to type your name, email address, and a password; and then tap Create Your Amazon Account.

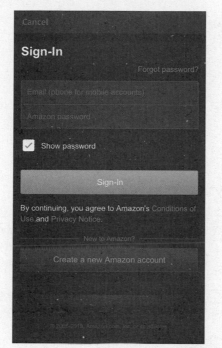

FIGURE 4-3:
The iPhone
version of the Fire
TV mobile app's
Sign In page.

TIP

2. **In the Amazon Password field, type your Amazon account's password.**

   By default, the Fire TV mobile app hides the password by displaying each character as a dot. If you want to make sure you entered the password correctly, select the Show Password check box.

3. **Tap the Sign In button.**

   The Fire TV mobile app confirms your Amazon credentials and then signs in to your account.

   If you've enabled two-factor authentication on your Amazon account (see Chapter 12), the Fire TV mobile app will prompt you to enter a one-time password (OTP) to verify the sign-in. Type the code that was sent to you and then tap the Sign In button.

   The Fire TV mobile app finishes loading, and you see the home screen, which will be similar to the one shown in Figure 4-4.

Hello Paul McFedries

Select a Device to Connect

Paul's Fire TV Stick >

Paul's TV >

+ Set up a Fire TV Recast (US only)

Sign Out

FIGURE 4-4:
The Fire TV
mobile app's
home screen.

Pairing your mobile device with Fire TV

The Fire TV mobile app will automatically detect your Fire TV device as long as both are on the same Wi-Fi network. In that case, the app's home screen shows your Fire TV device or devices, as shown earlier in Figure 4-4.

To use the app with a Fire TV device, you have to connect them by following these steps:

1. **Run the Fire TV mobile app and display the home screen.**

2. **Tap the Fire TV device you want to control.**

   The first time you do this, your Fire TV device displays a four-digit code, as shown in Figure 4-5.

   Your code will almost certainly be different from the one shown here. So, don't enter the code you see in this book; enter the code you see on your screen.

   REMEMBER

3. **In the Fire TV mobile app, type the four-digit code in the spaces provided, as shown in Figure 4-6.**

   Fire TV connects your mobile device with your Fire TV device.

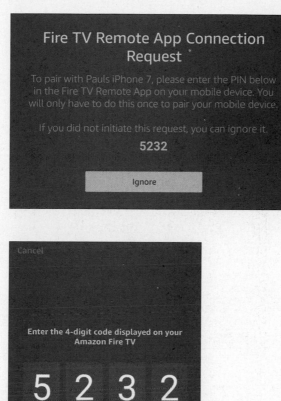

FIGURE 4-5:
Your Fire TV
device displays a
four-digit code.

FIGURE 4-6:
In the Fire TV
mobile app, enter
the four-digit
code to connect
the devices.

Touring the Fire TV mobile app remote

With your mobile device connected to your Fire TV device, you can control Fire TV using the Fire TV mobile app's Remote feature. When you use the Fire TV mobile app to select a Fire TV device, you see a Remote screen similar to the one shown in Figure 4-7. I say "similar to" for two reasons:

>> Figure 4-7 shows the Android smartphone version of the Remote screen. The tablet or iPhone version will have a slightly different look (although with the same layout).

>> Figure 4-7 shows the Remote screen for a Fire TV Edition device. The Remote screen for a Fire TV device doesn't include the Volume Up, Volume Down, and Mute buttons.

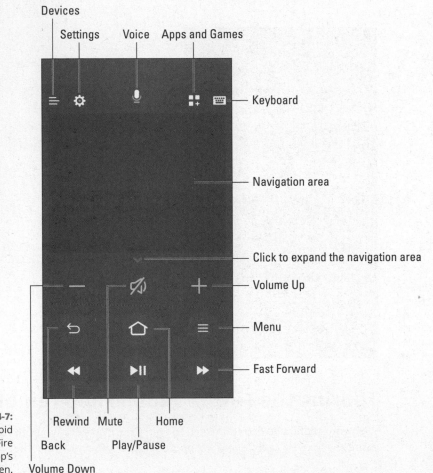

Devices

Settings   Voice   Apps and Games

Keyboard

Navigation area

Click to expand the navigation area

Volume Up

Menu

Fast Forward

FIGURE 4-7:
The Android
version of the Fire
TV mobile app's
Remote screen.

Rewind   Mute   Home

Back   Play/Pause

Volume Down

The buttons in the Fire TV mobile app's Remote screen are mostly identical to those on the Alexa Voice Remote, which I describe earlier in this chapter (see "Getting to Know the Alexa Voice Remote"). However, at the top of the Fire TV mobile app's Remote screen, there are four unique buttons:

>> **Devices:** Tap this button to return to the list of your Fire TV devices.

>> **Settings:** Tap this button and then tap Settings to display your Fire TV device's Settings screen. You can also tap this button and then tap Sleep to put your Fire TV device into sleep mode.

>> **Apps and Games:** Tap this button to display the Apps and Games screen, which displays an icon for each app and game installed on your Fire TV device.

>> **Keyboard:** Tap this button to display an onscreen keyboard for entering text on the Fire TV screen (for the details, see "Entering text with the Fire TV mobile app keyboard," later in this chapter).

I should also mention that the navigation area in the Fire TV mobile app doesn't work the same as the navigation ring on the Alexa Voice Remote. To learn what's different about the app's navigation area, see "Navigating with the Fire TV mobile app," later in this chapter.

Navigating the Fire TV Interface

Getting around the Fire TV interface can be a tad confusing at first. To help reduce the initial learning curve, remember that navigating the screen essentially means scrolling through two types of onscreen elements:

>> **Commands:** These are usually text, but they can also be icons (such as the magnifying glass that indicates the Fire TV Search feature). The commands you'll deal with most often are the items that run across the top of the Fire TV Home screen: Home, Live, Your Videos, and so on.

>> **Tiles:** These are rectangles that represent apps, TV features (such as inputs), and items in the channel guide.

In the next two sections, I describe how you navigate the Fire TV interface using the Alexa Voice Remote and the Fire TV mobile app. Before I get to that, however, you may be asking yourself, "How will I know 'where' I am on the screen as I navigate?"

Great question! As you move about the interface, Fire TV moves a *selector* that highlights the current item. The look of the selector depends on the type of screen element:

>> If the highlighted element is a command, Fire TV displays that command in orange text rather than the standard white text, as shown in Figure 4-8.

>> If the highlighted element is a tile, Fire TV expands the tile slightly and adds a border around the tile, as shown in Figure 4-9.

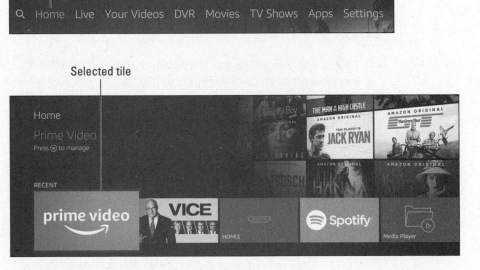

Selected command

FIGURE 4-8:
Fire TV displays a highlighted command in orange text.

Q   Home   Live   Your Videos   DVR   Movies   TV Shows   Apps   Settings

Selected tile

Home
Prime Video
Press ⊙ to manage

RECENT

FIGURE 4-9:
Fire TV displays a highlighted tile expanded with a border.

Navigating with the Alexa Voice Remote

With the Alexa Voice Remote, you mostly navigate the Fire TV interface using the navigation ring, shown in Figure 4-10.

As pointed out in Figure 4-10, there are four "buttons" on the ring:

>> **Up:** This is the top part of the ring; you press this button to move the selector up in the current Fire TV screen.

>> **Down:** This is the bottom part of the ring; you press this button to move the selector down in the current Fire TV screen.

» **Left:** This is the left part of the ring; you press this button to move the selector to the left in the current Fire TV screen.

» **Right:** This is the right part of the ring; you press this button to move the selector to the right in the current Fire TV screen.

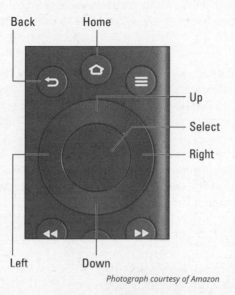

Back   Home

Up

Select

Right

Left   Down

FIGURE 4-10:
The Fire TV interface navigation buttons on the Alexa Voice Remote.

Photograph courtesy of Amazon

When the selector is on the item you want to work with, press the Select button (refer to Figure 4-10) to choose that item.

To return to the previous Fire TV screen, press the Back button; to return directly to the Home screen, press the Home button (refer to Figure 4-10).

Entering text with the Alexa Voice Remote

Most of your Fire TV navigation efforts will involve using the Alexa Voice Remote Up, Down, Left, and Right buttons to move the selector to the item you want to choose. More often than you may think, however, Fire TV asks you to enter text. It may be sign-up or sign-in data for a streaming service, a network password, or a search request.

With Fire TV, you can enter text the easy way or the hard way. The easy way depends on what type of text Fire TV is asking for:

» For a search request, it's easiest to use your voice to tell Alexa what you're looking for (see "Introducing Voice Control of Fire TV," later in this chapter).

» For all other text inputs, it's easiest to use the keyboard that comes with the Fire TV mobile app (see "Entering text with the Fire TV mobile app keyboard," later in this chapter).

What if you're in a situation where you can't use your voice (because, say, you don't want to disturb people nearby) and you don't have the Fire TV mobile app handy? In those cases, you need to enter your text the hard way, which means using the Alexa Voice Remote navigation ring.

The idea is that when Fire TV requires some text, it displays an onscreen keypad. The layout of that keypad depends on the type of text that Fire TV requires. For example, Figure 4-11 shows the onscreen keypad for a password, while Figure 4-12 shows the keypad for a search.

Typed characters appear here

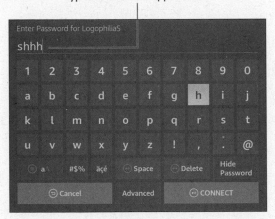

FIGURE 4-11:
The Fire TV onscreen keypad for entering a password.

Here's the basic procedure:

1. **Use the navigation ring's Up, Down, Left, and Right buttons to highlight the character you want to enter.**

As you move through the keypad, Fire TV moves the selector to show you the highlighted character.

2. **Press the Select button to choose the character.**

   Fire TV adds the character to the text box.

3. **Repeat Step 1 and 2 until you've entered all your text.**

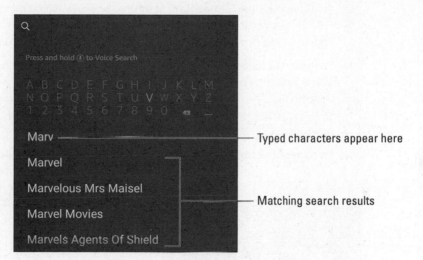

Typed characters appear here

Matching search results

FIGURE 4-12:
The Fire TV
onscreen keypad
for searching.

Yep, it's painful for all but the shortest text entries. Fortunately, Fire TV offers a few shortcuts that you can take advantage of:

>> Press Fast Forward to enter a space.

>> Press Rewind to delete the most recent character.

>> In keypads that support both lowercase and uppercase letters, press Menu to toggle the keyboard's letters between lowercase and uppercase.

>> In keypads where you have to confirm your text entry (such as the Connect button in Figure 4-11), press Play/Pause to choose the confirmation button.

>> In keypads that offer a Cancel button (again, refer to Figure 4-11), press Back to choose that button.

Navigating with the Fire TV mobile app

With the Fire TV mobile app, you mostly navigate the Fire TV interface using the navigation area. There are three techniques to master:

>> **To move the selector one item at a time,** briefly swipe inside the navigation area in the direction you want the selector to move. For example, to move the selector down one item, briefly swipe down in the navigation area.

>> **To move the selector quickly across multiple items,** tap and hold anywhere inside the navigation area to display the navigation ring; then drag your finger in the direction you want to move the selector. For example, to move the selector to the left, drag your finger to the left; the left portion of the navigation ring lights up, as shown in Figure 4-13.

>> **To choose the highlighted item,** tap anywhere within the navigation area.

FIGURE 4-13:
Tap and drag within the navigation area to move the selector (such as to the left, as shown here).

To return to the previous Fire TV screen, press the Back button; to return directly to the Home screen, press the Home button (refer to Figure 4-1).

Entering text with the Fire TV mobile app keyboard

Earlier I mention that using the Alexa Voice Remote navigation ring to type one character at a time is a slow and cumbersome way to enter text. Entering text via voice, as I describe in the next section, is certainly easy, but not always practical (say, if people are nearby) or possible (say, if you're entering a password).

Fortunately, Fire TV gives you a third method for entering text: the Fire TV mobile app's onscreen keyboard. This is a standard mobile device keyboard that you can use to quickly type whatever text you want to enter into the Fire TV interface. Here's how it works:

1. **In the Fire TV mobile app, connect to the Fire TV device you want to control.**

2. **Navigate to the Fire TV item that requires text input, such as the Search feature or the sign-in screen for a streaming app.**

   As soon as you navigate to a field that requires text, the Fire TV mobile app automatically displays the onscreen keyboard.

TIP

   If the onscreen keyboard doesn't appear, tap the Keyboard icon (refer to Figure 4-7).

3. **Use the keyboard to type your text entry.**

   The letters you type appear near the top of the app screen, as shown in Figure 4-14, as well as in the Fire TV text field on your TV.

Typed characters appear here

FIGURE 4-14: Use the Fire TV mobile app's onscreen keyboard to enter text into a Fire TV text field.

Introducing Voice Control of Fire TV

I discuss voice control of Fire TV in depth in Chapter 9. For now, let me give you a quick introduction to using Alexa to control your Fire TV.

The first thing you need to know is that there are actually *two* types of voice control for Fire TV:

>> **Near-field:** The microphone that picks up your voice must be close to you (say, no more than arm's length). This is the type of voice control at work when you use the Alexa Voice Remote that comes with your Fire TV device.

>> **Far-field:** The microphone that picks up your voice can be relatively far away — up to about 20 feet. This is the type of voice control at work when you incorporate an Alexa-enabled device, such as a Fire TV Cube or an Echo smart speaker, into your Fire TV system (see Chapter 9).

Issuing voice commands using the Alexa Voice Remote

For *utterances* (as Amazon likes to call the verbal directives you send Alexa's way) that fall under the near–field rubric, follow these steps:

1. **Press and hold the Voice button on the Alexa Voice Remote.**

    Fire TV displays a blue bar across the top of the screen to indicate that Alexa is listening, as shown in Figure 4-15.

2. **State what you want Alexa to do.**

3. **Release the Voice button.**

    Alexa carries out your request (or, at least, tries to).

FIGURE 4-15:
When you press and hold the Voice button, a blue bar appears at the top of the screen.

Blue bar

Issuing voice commands using the Fire TV mobile app

To issue a voice command using the Fire TV mobile app, follow these steps:

1. **In the Fire TV mobile app, connect to the Fire TV device you want to control.**

2. **Drag down the Voice button (refer to Figure 4-7) until you see the text "Listening . . ." on the screen (see Figure 4-16) and then hold your finger there.**

   The first time you drag down the Voice button, Fire TV asks your mobile device for permission to continue:

   - On iOS, Fire TV asks for permission to use the microphone. Tap OK to grant that permission.

   - On Android, Fire TV asks permission to record audio. Tap Allow to grant that permission.

3. **Let Alexa know what you want it to do.**

4. **Release the Voice button.**

   Alexa carries out your command.

FIGURE 4-16:
Drag down the Voice button to get Alexa listening for your command.

Learning some useful voice commands

I run through a long list of possible utterances in Chapter 9, but here are a few of the most useful ones to get you started:

>> "Find [*title*]" (for example, "Find *Star Trek*").

>> "Find [*title*] [*type*]" (for example, "Find *Star Trek* TV show").

>> "Find [*genre*]" (for example, "Find comedies").

» "Play [*title*]" (for example, "Play *Fleabag*").

» "Play [*title*] [*type*]" (for example, "Play *M*A*S*H* movie").

» "Fast forward."

» "Rewind."

» "Pause."

» "Play" (or "Resume").

» "Turn volume up/down."

» "Mute."

REMEMBER

If you're speaking these commands to a Fire TV Cube, Echo, or similar Alexa-enabled device, be sure to precede the utterance with the word *Alexa*.

Looking Around the Fire TV Home Screen

When you start up your Fire TV Edition Smart TV, or when you switch to your Fire TV input, you come face-to-face with the Fire TV Home screen, which looks something like the one shown in Figure 4-17.

Featured

Tabs

FIGURE 4-17:
An example of a
Fire TV Home
screen.

Recent

Across the top of the screen are a series of tabs that take you to different areas of the Fire TV interface. Here's a quick summary of what each tab represents:

>> **Search (magnifying glass icon):** Enables you to search the Fire TV interface for apps, movies, TV shows, and other content.

>> **Home:** Displays the Fire TV Home screen.

>> **Live:** Displays all your live TV sources, including apps that offer live TV (such as Hulu + Live TV and Pluto TV), Fire TV Edition antenna or cable channels, and Fire TV Recast over-the-air channels.

>> **Your Videos:** Displays movies and TV shows that you're watching (or have watched) and your Prime Video Watchlist of shows you'd like to see.

>> **DVR:** Shows current and recorded over-the-air TV shows. Note that you only see this tab if you have a Fire TV Recast device connected to your network (see Chapter 7).

>> **Movies:** Displays the movies available through your subscription services.

>> **TV Shows:** Displays the TV shows available through your subscription services.

>> **Apps:** Displays featured apps for streaming services and other content.

>> **Settings:** Displays the settings for customizing and configuring Fire TV.

Below the tabs, you see the Featured section, which displays content that Amazon wants to highlight. Fire TV describes the Featured section as a *rotator* because when you navigate into the Featured section, it begins automatically rotating through the content. You can also navigate the featured content manually by pressing the navigation ring's Right and Left buttons. To get out of the previews, press the Back button.

Below the Featured section, Fire TV displays a series of rows, each of which is populated with a bunch of tiles, mostly for streaming apps. The first row is always Recent, which displays your eight most recently viewed apps.

To remove an item from the Recent row, highlight it, press the Menu button, and then choose Remove from Recent.

Other rows include Your Apps & Games, Inputs (Fire TV Edition Smart TVs only), Top Movies, Top TV, and Top Free Games.

Chapter 5

Watching Live TV

Through free streaming services such as IMDb TV and YouTube, subscription streaming services such as Amazon Prime Video and Netflix, and web-based videos available through Fire TV browsers such as Amazon Silk and Firefox, you have a huge collection of movies, TV shows, and other video content at your disposal. Not, however, a *complete* collection. What's missing? The biggest gap in your Fire TV video-watching toolkit is the absence of live TV, which refers to content that can only be watched at a time predetermined by the broadcaster. (In contrast to *on-demand* content, which you can watch any old time you feel like it.) Live TV generally refers to two types of programming:

» Prerecorded shows broadcast at specific times

» Live events, such as sports and breaking news

It's quite possible you can exist happily without live TV. However, if you're a sports nut, a news junkie, or someone who simply must watch *Saturday Night Live* when it's actually *live*, then you need to augment your Fire TV system with live TV. There are two routes to live TV you can take:

» Connect an HDTV antenna to your Fire TV Edition device.

» Install a live TV app on any Fire TV device.

This chapter takes you through these live TV scenarios and shows you how to watch and manage your live TV channels.

There's a third live TV scenario: If you have an existing cable set-top box (or satellite receiver), connect that device to the appropriate input jack on your TV; then switch to that input.

There's even a fourth live TV scenario: Add a Fire TV Recast to your network and connect an HDTV antenna to the Recast to both watch and record live over-the-air channels. It's a sweet setup, and it's the subject of Chapter 7.

Getting Live TV through an Antenna

Perhaps the simplest — and certainly the cheapest — way to do live TV is to purchase an HDTV antenna and connect it to your TV. This works because TV stations around the country — including network affiliates and independent stations — broadcast their live TV signals into the ether. These so-called *over-the-air* (OTA) signals can be picked up by an HDTV antenna.

The signals you can pick up depend on various factors, including your location (the closer you are to broadcasting stations, the more likely you are to pick up the signals) and your antenna (generally speaking, smaller indoor antennas have less range than larger outdoor antennas). See Chapter 10 for more details on signal pickup and how to determine what signals are available in your area.

If you want to access your antenna's channels through the Fire TV interface, then you need to be using a Fire TV Edition device (such as a Fire TV Edition Smart TV) because only Fire TV Edition comes with a built-in TV tuner. Other Fire TV devices such as the Fire TV Stick 4K and the Fire TV Cube don't come with a tuner, so they can't do the live TV thing.

Connecting your antenna

Your antenna comes with a coaxial cable that you attach to the corresponding port on your TV, which is usually labelled *Antenna* (or just *Ant*) and/or *Cable* (or *Cable In*), as shown in Figure 5-1.

Now position the antenna. For a rooftop or attic antenna, you're probably getting the antenna mounted by a professional. For a simpler indoor antenna, mount the device on or close to the nearest window and try to angle the antenna toward the broadcast signals you most want to pick up. (Again, see Chapter 10 to learn how to figure out which directions your available signals are coming from.)

HDTV antenna (indoor)

Antenna port

Coaxial cable

FIGURE 5-1:
Connect
your antenna's
coaxial cable to
your TV's Antenna
or Cable port.

Scanning for channels

With your antenna connected, your next chore is to ask Fire TV to scan for the available channels. Here are the steps to follow:

1. One your Fire TV Edition device, choose Settings ⇨ Live TV.

If you don't see the Live TV setting, then it means you're using a Fire TV device which, as I mention earlier, doesn't support live TV. To see the Live TV setting, you must be using a Fire TV Edition device.

2. Choose Channel Scan.

Fire TV tells you to make sure your antenna is connected and positioned where you want it.

3. You've done all that, so say "Check!" and choose Next.

Fire TV begins scanning the antenna connection for signals and displays its progress, as shown in Figure 5-2.

When the scan is complete, Fire TV shows you how many channels it found, as shown in Figure 5-3.

TIP

If Fire TV found far fewer channels than you expected, reposition the antenna and then choose Rescan to try again.

4. Choose Done.

Your live TV channels are ready to view.

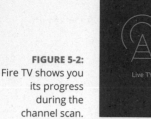

FIGURE 5-2:
Fire TV shows you its progress during the channel scan.

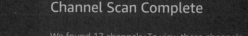

Scanning for channels...

Channel scan is 50% complete.
Channels found: 13
Currently scanning for digital channels only.

Live TV

Cancel

FIGURE 5-3:
When the scan is complete, Fire TV shows you how many channels it found.

Channel Scan Complete

We found 17 channels. To view these channels, go to the Home tab.

Live TV

Done   Rescan   Channel Management

Checking channel signal strength

As shown in Figure 5-3, when you scan for Live TV channels, Fire TV tells you how many channels it found, but it says nothing about the strength of each channel's signal. Sure, you're free to tweak the position of your antenna and rescan to get more channels, but all you're doing is maxing out the number of channels. You may prefer to get the best signal you can for those channels you actually watch regularly.

Fortunately, Fire TV can tell you the signal strength of each channel that it found, which means you can check the signal for each channel that you'll be watching frequently. Here are the steps to follow:

1. On your Fire TV Edition device, choose Settings ⇨ Live TV ⇨ Channel Management ⇨ Antenna Channels.

Fire TV displays the Antenna Channels screen and selects the All tab, which shows a list of the channels Fire TV found during the scan.

REMEMBER

To navigate the All, Favorites, and Hidden tabs, press the Left and Right buttons on the Alexa Voice Remote navigation ring.

2. Select the channel you want to check.

On the right side of the screen, Fire TV displays a preview of the channel and the Signal Strength value: good, poor, or unavailable. Figure 5-4 shows an example.

3. If the channel preview is bad quality and/or the Signal Strength value is poor, try repositioning the HDTV antenna until you get a good signal.

4. Repeat steps 2 and 3 for the other channels you want to check.

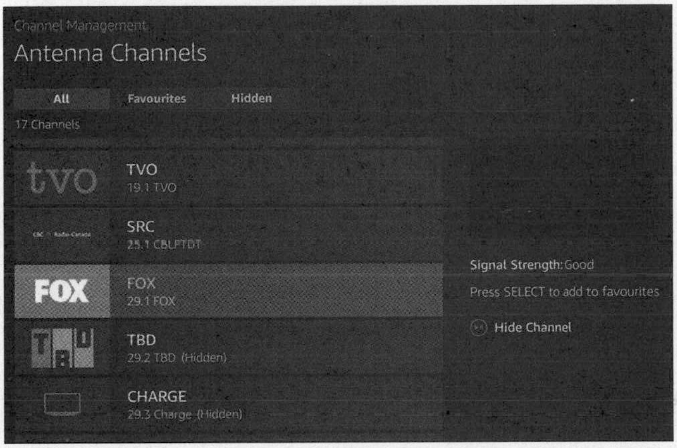

FIGURE 5-4:
In the Antenna Channels list, select a channel to see its Signal Strength value.

TIP

If your antenna channels appear distorted or otherwise not as nice to look at as you'd like, Fire TV Edition offers several settings that can help. To adjust the picture, choose Settings⇨Display & Sounds⇨Picture Settings⇨Antenna. Fire TV switches to a live TV channel and displays the Picture settings, including Picture Mode, Contrast, and Brightness. To adjust the display, choose Settings⇨Display & Sounds⇨Display Settings⇨Antenna. Fire TV switches to a live TV channel and displays the Display settings, which include the display aspect ratio.

Getting Live TV Using a Third-Party App

Getting live TV through an HDTV antenna hookup is free, but you usually only get a few stations and you don't get premium cable stations such as HBO and Showtime. If you want to improve your live TV experience with more channels and better content, and you don't mind spending a few extra dollars each month, then you need to investigate the world of third-party Fire TV apps that offer so-called *over-the-top* (OTT) live TV feeds.

REMEMBER

An over-the-top TV service is one that's delivered via a broadband Internet connection. It's called "over-the-top" because, in a sense, such services "jump over" your existing cable box or satellite receiver and give you content directly.

A few free services offer live TV, but in most cases you need a subscription (although almost all live TV app providers offer a free trial period). Here are a few third-party apps that provide live TV channels (among other offerings, in most cases):

>> **AT&T TV Now:** www.atttvnow.com

>> **fuboTV:** www.fubo.tv

>> **Hulu with Live TV:** www.hulu.com/start/live-tv

>> **Philo:** https://try.philo.com/

>> **Pluto TV:** https://pluto.tv

>> **Sling TV:** www.sling.com

>> **YouTube TV:** https://tv.youtube.com

You need an account to use any of these options, so before installing a service's Fire TV app, go to the service's website to sign up for an account.

WARNING

Many of the services listed here are not available in all countries; most are available only in the United States. Frustrating! Use the Fire TV Search feature to search for "live" to see what apps are available in your area.

Managing Live TV channels

If you're itching to watch something live, don't let me stop you: Grab your favorite snack and head directly to this chapter's "Watching Live TV" section.

When you're done, come back here to learn some important Fire TV channel management features that will make watching Live TV an easier and more efficient experience.

Adding a live TV channel to your favorites

As I discuss later (see "Navigating the live TV channel guide"), Fire TV creates a channel guide to help you navigate all your live TV channels. That guide has a section for your antenna channels, as well as a section for each live TV app you've installed (plus yet another section for your Fire TV Recast channels; see Chapter 7). That's a lot of sections, so locating the channel you want can be a hassle.

Yep, you can quickly tune directly to any channel using a voice instruction (see Chapter 9), but voice isn't always convenient or possible. Another way to make the channel guide more efficient is to mark one or more channels as favorites, which then handily appear at the top of the channel guide (in a section labeled Favorite Channels; see Figure 5-5) for easy access.

FIGURE 5-5: In the channel guide, live TV channels marked as favorites appear in the Favorite Channels section.

On your Fire TV Edition device, there are two methods you can use to mark a channel as a favorite:

>> Choose Settings ➪ Live TV ➪ Channel Management, choose your live TV source (such as Antenna Channels), highlight the channel you want to work with, and then press the Select button to add the channel to the favorites list.

To use Settings to remove a channel from Favorites, highlight it either on the All tab or on the Favorites tab (see Figure 5-4), and then press the Select button.

>> Highlight the channel you want to work on the Home tab (in the Recent row or in the On Now row), the Live tab, or the channel guide, press the Menu button on your Alexa Voice Remote, and then choose Add to Favorite Channels in the shortcut menu that appears.

To remove a channel from Favorites, highlight it in the channel guide's Favorite Channels section (or elsewhere in the Fire TV interface), press the Menu button, and then choose Remove from Favorite Channels in the shortcut menu.

Hiding a live TV channel or app

If there are live TV stations that you never watch, you can make the channel guide easier and quicker to navigate by hiding those unwatched stations. On your Fire TV Edition device, there are two methods you can use:

REMEMBER

» Choose Settings ➪ Live TV ➪ Channel Management, choose your live TV source (such as Antenna Channels), highlight the channel you want to work with, and then press the Play/Pause button.

To show a hidden channel, either highlight the channel in the All tab and press Play/Pause, or highlight the channel in the Hidden tab (see Figure 5-4) and press Select.

» Highlight the channel you want to hide on the Home tab (in the Recent row or in the On Now row), the Live tab, or the channel guide, press the Menu button on your Alexa Voice Remote, and then choose Hide Channel in the shortcut menu that appears.

Filtering live TV channels

Depending on what you have connected to your TV and which live TV apps you have installed, your live TV lineup may be quite long, making the channel guide cumbersome to navigate. Chances are, there are lots of channels you never view, so one solution would be to hide those channels, as I describe in the preceding section.

Hiding channels works fine, but it can be a time-consuming chore if you have lots of channels you want to hide. An often better solution is to filter the channel guide to show only those channels from one of the following channel types:

» **Sources:** Shows only those channels that you receive from a particular live TV source, such as Antenna Channels or Fire TV Recast Channels.

» **Apps:** Shows only those channels that are available through a particular app.

» **Favorites:** Shows only those channels that you've added to your Favorites list.

Follow these steps on your Fire TV device to filter the channel guide:

1. **Display the channel guide.**

    See "Navigating the live TV channel guide," later in this chapter.

2. **On your Fire TV remote, press Menu.**

    Fire TV displays the Options menu for the channel guide.

3. **Choose Filter Channels.**

Fire TV displays the Filter menu, which contains a list of possible filters. Figure 5-6 shows an example.

4. **Choose the filter you want to use.**

Fire TV updates the channel guide to show only the channels in the filter you chose.

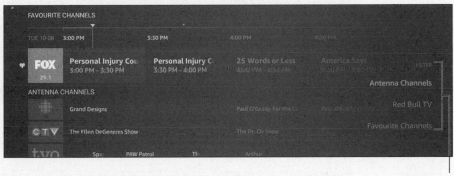

FIGURE 5-6: In the channel guide, choose Menu ⇨ Filter to see a list of filters you can apply.

Filter menu

Watching Live TV

With a source such as an HDTV antenna attached to your TV, or one or more third-party live TV apps installed, you're ready to start watching live TV. The rest of this chapter takes you through several ways to watch live stuff on your Fire TV Edition device.

Seeing what's on now

When your Fire TV device has one or more live TV services, the Home screen sprouts a new row titled On Now, shown in Figure 5-7. The On Now row shows thumbnail views of the current live TV shows. Use your Fire TV remote's navigation ring to scroll right and left through the shows in the On Now list. For each show you land on, Fire TV displays the name, start and stop times, and a brief description. When you land on a show you want to watch, press the Select button.

To return to the On Now list, you have two ways to go:

>> If you want to return to the Home screen's On Now row, press the Back button on the Fire TV remote.

>> If you want to keep watching the current show, press the Down button on your Fire TV remote's navigation ring to display the On Now shows at the bottom of the screen. Press Up on the navigation ring to hide the On Now shows.

FIGURE 5-7:
In the Home screen, the On Now row shows you what's playing live right now.

Navigating the live TV channel guide

The On Now row that I discuss in the preceding section is handy, but what if you want to know not only what's on right now, but also what's coming up later today, tomorrow, or next week? To peer into the future of your live TV sources, you need to use Fire TV's channel guide. To get there, you have three choices:

>> On the Home screen, navigate down to the On Now row; then press the Left button on the navigation ring to reveal the Guide button, shown in Figure 5-8, and choose Guide.

>> Highlight any show in the On Now row, press the Fire TV remote's Menu button, and then choose Channel Guide.

>> On your Fire TV Edition remote, press the Channel Guide button.

The channel guide (see Figure 5-9) is divided into sections. If you've added one or more channels to your Favorites list, you see the Favorite Channels section at the top of the guide. You also see sections for each live TV source (such as Antenna Channels), and for each live TV app.

FIGURE 5-8:
The Guide button
resides to the left
of the leftmost
tile in the On
Now row.

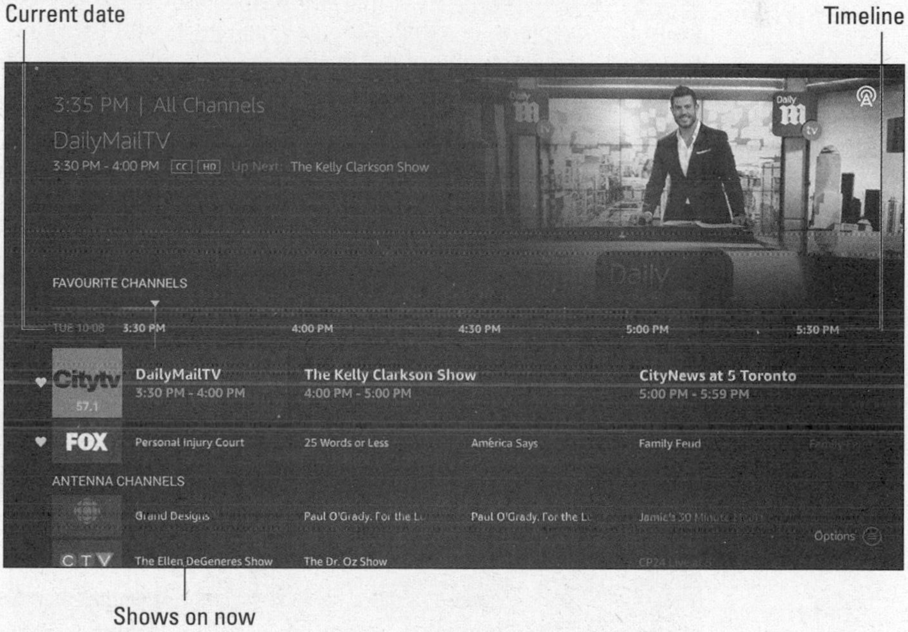

Guide button

Current date   Timeline

3:35 PM | All Channels

DailyMailTV

3:30 PM - 4:00 PM CC HD Up Next: The Kelly Clarkson Show

FAVOURITE CHANNELS

TUE 10-08   3:30 PM   4:00 PM   4:30 PM   5:00 PM   5:30 PM

Citytv DailyMailTV   The Kelly Clarkson Show   CityNews at 5 Toronto
57.1   3:30 PM - 4:00 PM   4:00 PM - 5:00 PM   5:00 PM - 5:59 PM

♥ **FOX**   Personal Injury Court   25 Words or Less   América Says   Family Feud

ANTENNA CHANNELS

         Grand Designs   Paul O'Grady: For the L...   Paul O'Grady: For the L...   Jamie's 30 Minute...

                                                                                            Options

CTV   The Ellen DeGeneres Show   The Dr. Oz Show

FIGURE 5-9:
The channel
guide is divided
into sections.

Shows on now

The channel guide grid displays the current date and a timeline. The channels themselves display a name and a logo on the far left, followed by the currently airing show. Here are the techniques you can use to navigate the channel guide:

>> Use your Fire TV remote's Down and Up navigation buttons to scroll vertically through the live shows; then press Select to tune to the show you want to watch. To return to the channel guide, press the Back button on the Fire TV remote.

>> Use your Fire TV remote's Right and Left navigation buttons to scroll horizontally through the upcoming shows. You can press Select to learn more about any show. To return to the channel guide, press the Back button on the Fire TV remote.

TIP

On your Fire TV remote, press Fast Forward to skip ahead one day in the channel guide, and press Rewind to skip back one day.

Checking out the Live tab

As an alternative to the channel guide, you can select the Live tab in the Fire TV Home menu bar. The Live screen that appears (see Figure 5-10) contains a Recent Channels row, which is a list of the live TV channels you've viewed most recently. Below that you see a row with some suggested live TV apps and rows that integrate live events under the headings Live Sports and Breaking News. The rest of the Live screen contains rows for your live TV sources (such as Antenna Channels), Fire TV Recast OTA shows, and any live TV apps you've installed.

FIGURE 5-10:
A typical Live
screen.

REMEMBER

As this book went to press, the Live tab was available only to Fire TV customers in the United States and Germany. When asked, Amazon said the Live tab will be coming to other countries in the near future.

Controlling live TV playback

Your Fire TV Edition Smart TV isn't a digital video recorder (DVR), but it does offer several DVR-like features that can improve your live TV viewing experience:

>> **Pausing live TV:** Press the Play/Pause button to pause the live playback. You can pause for up to two minutes before Fire TV Edition automatically resumes the show.

>> **Rewinding live TV:** Press Rewind to go back ten seconds. Press and hold Rewind to go back up to two minutes at faster speeds.

>> **Return to the earliest position:** Fire TV Edition maintains a two-minute buffer, meaning you can jump back two minutes at any time during playback. To jump back, press Menu and then choose Watch from Earliest.

TIP

To get a larger rewind buffer, connect an external storage drive to your Fire TV Edition Smart TV (see Chapter 8).

>> **Return to the live show:** If you've paused or rewound the show or jumped back to the earliest position, you can return to the live feed by pressing Menu and then choosing Jump to Live.

IN THIS CHAPTER

» **Installing apps for movie and TV streaming services**

» **Searching Fire TV for movies and TV shows**

» **Taking a look at some free streaming services**

» **Controlling movie and TV show playback**

» **Adding a second screen to your streaming experience**

Chapter **6**

Streaming Movies and TV Shows

F ire TV sure has a ton of tricks up its sleeves, and I take you through all of them — such as playing music, viewing photos, and surfing the web — later in the book. However, when you get right down to it, you probably bought your Fire TV device to do just two things — watch movies and TV shows — am I right? I knew it. Nothing wrong with that, of course, because Fire TV excels at streaming video content.

So, this chapter is devoted exclusively to using Fire TV to watch movies and TV show episodes. You learn how to install streaming apps, search for content, handle subscriptions, control playback, and much more. I'll wait here while you pop some popcorn, and then we'll get the show started.

Installing a Streaming Media App

Although your Fire TV comes with a few Amazon apps preinstalled — particularly Amazon Prime Video, Amazon Photos, and Amazon Music — before you can watch or even check out any other streaming content, you need to install the appropriate app on your Fire TV device.

Here's the procedure to run through to install an app on Fire TV:

1. **From the Fire TV Home screen, choose the Apps tab.**

2. **Choose the app you want to install.**

3. **Choose the Get button, shown in Figure 6-1.**

   Choosing Get authorizes Amazon to "purchase" the app through your Amazon account. All movie and TV apps are free to download, however, so you won't be charged anything.

   Note, too, that if you've previously purchased a Fire TV app (say, using a different Fire TV device), then you see a Download button instead of the Get button.

   Fire TV downloads and then installs the app. When the install is complete, you see the Open button, as shown in Figure 6-2. In the lower-right corner of the screen, you also see a notification that the app is ready to launch (again, see Figure 6-2; note that this notification only appears onscreen for a few seconds).

4. **Choose Open or, if the notification is still onscreen, press Menu.**

   Fire TV runs the app.

FIGURE 6-1:
Open the app
and choose Get
to download it.

FIGURE 6-2:
You see the Open button and the Ready to Launch notification after the app installation is complete.

Searching for Movies and TV Shows

After you settle in to the Fire TV experience, it doesn't take long before you realize there is a *ton* of stuff to watch! If you're an Amazon Prime member, you've got hundreds of TV shows and thousands of movies to check out. And the more streaming apps you install, the more content is at the tip of your fingers (or the tip of your tongue). Just installing YouTube alone gives you access to several billion (that's right, I said *billion*) videos, mostly of cats.

When you realize just how much content you have at your disposal, the very next realization is usually, "How in the name of Jeff Bezos am I supposed to find what I want?" Fire TV helps you out by displaying your most recently used apps in the Recent row of the Home screen, but many people new to Fire TV complain that they seem to spend great chunks of their precious leisure time scrolling through the Fire TV interface to find the app, show, or movie they want.

Fortunately, there's a better way: Fire TV offers three different techniques to search for the content you want to watch:

>> **Amazon catalog search:** This is the general Fire TV search, and it looks through all the content that's part of the Amazon catalog of movies, TV shows, apps, and games. This catalog includes not only Amazon content, but also content from third-party apps that have implemented Amazon's Universal Search feature.

>> **Fire TV in-app search:** This is a search that Fire TV runs on the content in a specific app. As I write this, Fire TV can only search within three apps: Amazon Music (see Chapter 8), Silk (Amazon's web browser; again, see Chapter 8), and YouTube.

>> **Third-party in-app search:** This is a search feature that a streaming service has added to its Fire TV app.

Searching the Amazon catalog

To perform a global search of the Amazon catalog on Fire TV, press Home to return to the Fire TV Home screen (if you're not there already), and then choose the Search tab (the magnifying glass icon to the left of the Home tab; see Figure 6-3).

FIGURE 6-3:
Open the Fire TV
Search screen by
choosing the
Search tab.

Search

Q Home   Live   Your Videos   DVR   Movies   TV Shows   Apps   Settings

After you have the Search screen displayed, you have three ways to initiate a search:

>> **The hard way:** Use the Alexa Voice Remote navigation ring to choose each character of your search text using the onscreen keypad.

>> **The easier way:** With the Fire TV mobile app connected to your Fire TV device, swipe down in the navigation area to move the selector into the onscreen keypad, which automatically displays the Fire TV mobile app keyboard. Now type your search request.

>> **The easiest way:** Press and hold the Voice button on the Alexa Voice Remote (or drag down and hold the Voice button in the Fire TV mobile app; or, if you have Fire TV Cube or a similar Alexa-enabled device, just say "Alexa") and then tell Alexa what you want to find.

REMEMBER

In the search requests that follow, remember that if you're using a Fire TV Cube or other Alexa-enabled device, be sure to precede the search request with the wake word *Alexa.*

Whatever method you prefer, the trick is to use some combination of keywords that tells Fire TV what you're looking for. There are five main types of keywords you can use:

>> "[*app*]" (for example, "YouTube")

>> "[*content*]" (for example, "cats")

>> "[*title*]" (for example, "*Star Trek*")

>> "[*type*]" (for example, "TV show")

>> "[*genre*]" (for example, "science fiction")

Some searches require only one keyword, but most of the time you'll want to combine two or more keywords to narrow your search. Here are some examples:

>> "[*title*] [*type*]" (for example, "*Star Trek* TV show")

>> "[*content*] [*genre*] [*content*]" (for example, "music documentary *Queen*")

The secret to efficient searching is to make your searches specific enough that you find what you're looking for, but not so wordy that Fire TV shrugs its shoulders and gives up. For example, this search doesn't work very well:

"The fourth *Star Trek* movie"

However, this pithier version works just fine:

"*Star Trek* Four"

If you run a voice search and Fire TV isn't certain what you said, it displays a list of possible searches, as shown in Figure 6-4. Choose the search text you want Fire TV to run. If you don't see the search request you want, press Menu to type the search text instead.

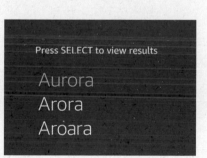

FIGURE 6-4:
If Fire TV isn't quite sure what you said, it displays a list of possible searches.

TIP

You don't have to navigate to the Search screen to run a voice search within Fire TV. Instead, from any Fire TV screen, begin your search request with the word *Find* (for example, "Find *Seinfeld*").

One final note about voice searching using Fire TV Cube (or an Alexa-enabled smart speaker): When Alexa returns multiple results, those results are numbered, as shown in Figure 6-5. You can use the following voice commands to navigate this list:

» **Scroll the list to the right:** "Alexa, scroll right" (or "Alexa, show more" or "Alexa, next").

» **Scroll the list to the left:** "Alexa, scroll left" (or "Alexa, previous").

» **Choose a result:** "Alexa, choose *X*," where *X* is the number of the item you want. Instead of "choose," you can say "select," "open," or "play."

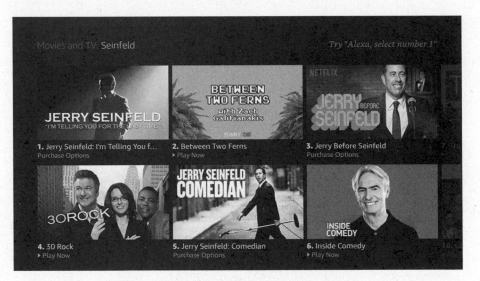

FIGURE 6-5:
With a Fire TV Cube search, Alexa displays a numbered list like the one shown here.

REMEMBER

In the search results, if you see "Purchase Options" below the result, it means the app requires a subscription before you can view the content.

Searching within an app using Fire TV

In the previous section, I mention several keyword types you can use for searching, including the name of an app, the title of a show, and a content description. I also mentioned that you can combine keywords, so you may think you can perform a search like this:

"[*app*] [*content*]" (For example, "Netflix dogs")

That command works *if* the app has added its content to the Amazon Universal Search catalog. Otherwise, the command fails because Fire TV doesn't know how to search within the app. I explain how to search within apps in the next section, but for now I can tell you that there are (at least as I write this) three apps that Fire TV *does* know how to search: Amazon Music, Silk (Amazon's Fire TV web browser), and YouTube.

REMEMBER

To conduct a Fire TV search within Amazon Music, Silk, or YouTube, you must first install the app on your Fire TV device.

To send a search request to one of these apps, run a global Fire TV search as I describe in the previous section. In the search results, you see a row titled Search in Apps, which includes a tile each for YouTube, Internet (that's the Silk browser), and Amazon Music, as shown in Figure 6-6. Choose one of these apps, and Fire TV passes along your search request to the app.

FIGURE 6-6:
Use the Search in Apps row to pass along your search text to one of the apps shown.

TIP

By default, when you pass along some search text to Silk, the browser uses the Bing search engine to scour the web for your search text. To use a different search engine (such as Google), open Silk, press Menu, scroll to the bottom of the screen and choose Settings ⇨ Advanced ⇨ Search Engine. Select the search engine you want to use for web searches.

You can also use voice commands to search Amazon Music and YouTube (for some reason, Silk doesn't respond to voice commands):

>> "Find [*video*] in YouTube," where *video* is a video title or keyword (for example, "Find cats in YouTube").

>> "Find [*music*] in Amazon Music," where *music* is a genre or the name of a song, album, or artist (for example, "Find salsa in Amazon Music").

Using an app's search feature

I mention in the preceding section that Fire TV can't search within most apps (with Amazon Music, Silk, and YouTube being the exceptions). That's not a huge problem, however, because most streaming apps implement their own search feature. For example, Figure 6-7 shows the search interface for the YouTube app.

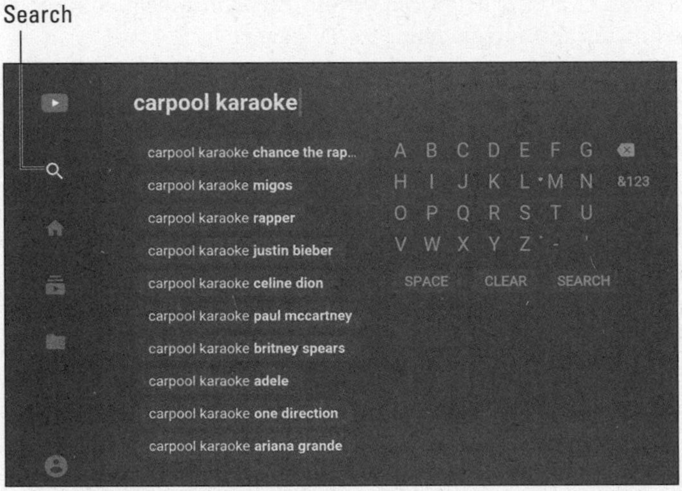

Search

FIGURE 6-7:
The in-app search feature of the YouTube app.

That's the good news. The bad news is that, just to keep us all on our toes, each app's search feature is slightly different, so there's no standard procedure I can give you. Fortunately, almost all apps do implement two search techniques:

>> The search feature is marked by a magnifying glass icon (shown on the left side of the screen in Figure 6-7), similar to one in the Fire TV menu bar.

>> You enter your search text using an onscreen keypad (shown on the right side of the screen in Figure 6-7), which means you can also enter the text using the Fire TV mobile app's keyboard.

Streaming on the Cheap with Free Services

It's most unfortunate that a significant portion of people who are new to Fire TV believe that after they've purchased a Fire TV device, the entire world of streaming media is available to them without incurring any extra charges. Would that it were true! So, I hope I'm not dashing your hopes too cruelly when I tell you that most streaming services require a paid subscription.

TIP

Ah, but take note of that little word *most*. Happily, a few streaming services offer free TV shows and movies. Sweet! To help you get your TV and movie streaming career off to a good start, here's a list of a few free services to check out:

>> **Amazon Prime Video:** Well, okay, this one's only "free" if you already have a Prime membership. However, if you're a paid-up member of Amazon Prime, then look out, because you get free access to a massive catalog of TV shows and movies, including lots of original Amazon content, such as award-winning shows like *The Marvelous Mrs. Maisel* and *Fleabag.* The Fire TV Home screen even has a row dedicated to original Amazon programming that broadcasts in 4K Ultra-High Definition (if your TV supports that format).

>> **IMDb TV:** Install the IMDb TV app to view popular TV shows and blockbuster Hollywood movies, all free because IMDb TV is supported by advertising. Note that, as I write this, IMDb TV is only available to residents of the United States.

>> **Tubi:** Install the Tubi app to get access to thousands of TV shows and movies. Yep, it's all completely free, but you'll have to suffer through, er, I mean, *watch* a few ads along the way.

>> **YouTube:** Install the Fire TV YouTube app to get immediate access to the kazillions of videos that YouTube offers. As I mention in the preceding section, Fire TV knows how to search within the YouTube app, so any video you need is but a quick search away.

Subscribing to TV and Movie Streaming Services

You may find everything you need to be a happy couch potato in Amazon Prime Video, YouTube, and the other free streaming services I mention in the previous section. However, it's more likely that there will be premium cable channels (such as HBO and Showtime) and live news and sports channels that reside in your "Must Have" list. That means you're going to have to pay a monthly subscription to one or more streaming services to get the content you want.

How you subscribe depends on how you navigate to the content. For example, if you install and then open an app that requires a subscription, you usually see a Subscribe button, as shown in Figure 6-8. If subscription content comes up in a Fire TV search, then you usually see a Watch with [*Service*] 30-Day Free Trial button (where [*Service*] is the name of the streaming service), as shown in Figure 6-9.

FIGURE 6-8:
Subscription
services usually
offer a Subscribe
button when you
start the app.

ACORN TV

Not a member? | Already have an account? | Preview our shows

Subscribe | Log In | Browse

FIGURE 6-9:
Most streaming
services offer a
trial period to
watch content.

Rumpole of the Bailey

IMDb 8.2/10 NR CC

Season 1 Cigar puffing, whiskey sipping, insult-generating defense
barrister Horace Rumpole (played by Leo McKern) proves why he's one
of the most unexpected successful attorneys in Great Britain in this ...

Starring: Leo McKern, Jonathan Coy, Julian Curry
Audio: English Subtitles: English

Watch with Acorn TV
30-Day Free Trial | Seasons &
Episodes | Add to
Watchlist

However, instead of relying on the Fire TV interface to initiate your **subscription**,
it's usually easier and faster to go to the streaming service's website and set up
your subscription there. Be sure to sign up for whatever free trial is offered, just
to make sure you like the service before paying for it.

REMEMBER

Some streaming apps conveniently enable you to set up a subscription using your
Amazon billing info (as long as you have a 1-Click payment method set up for your
Amazon account). If you subscribe to a streaming service via Amazon, you won't be
able to manage that subscription on the streaming service website. Instead, sign in to
your Amazon account, choose Account & Lists⇨Your Account⇨Apps and More⇨
Your Subscriptions (or surf directly to www.amazon.com/appstoresubscriptions).

Buying or Renting a Movie or TV Show

One of the major benefits of being an Amazon Prime member is that you get access to tons of movies and TV shows at no extra charge through the Amazon Prime Video app. It's a binge-watcher's paradise. If, on the other hand, you're not a member of Amazon Prime, you can still access all those shows, but you need to either buy or rent them on your Fire TV device using Amazon Instant Video.

Here are the steps to follow to buy or rent a movie or TV show:

1. **Locate the movie or TV shows you want to buy or rent.**

2. **Choose the movie or TV show to display its product details.**

3. **Choose either the Buy button or the Rent button.**

   Amazon uses your 1-Click payment method to charge the rental or purchase to your account.

4. **To watch the content, you have two choices:**

   - To watch the content right away, choose the Watch Now button.

   - To watch the content later, navigate to the Fire TV Your Videos tab, and then open the content from the Your Videos screen.

Watching a Movie or TV Show

When it's time to watch a movie or TV show, you first need to locate the content you want to view. Fire TV gives you a bunch of options:

- ⟫ Search for the movie or TV show, as I describe earlier in the "Searching for Movies and TV Shows" section.

- ⟫ Resume (or rewatch) recent content by navigating to the Home screen's Recent row.

- ⟫ Choose the Your Videos tab, and then use the Watchlist row to access content you've added to your Amazon Watchlist. Your Videos also includes recommendations from subscription services.

- ⟫ Choose the Movies tab to see featured and recommended movies from Amazon and your subscription movie services.

- ⟫ Choose the TV Shows tab to see featured and recommended TV shows from Amazon and your subscription TV services.

When you choose a movie or TV show, you first see a details screen that describes the content. Figure 6-10 shows a typical details screen for Amazon content, which offers the following controls:

>> **Watch Now with [*Service*]:** Choose this button to start watching the content (where [*Service*] is the name of the streaming service that offers the content). If the content is a movie, it starts right away; if the content is a TV show, Fire TV begins playing the first episode of the first season.

>> **Seasons & Episodes:** For a TV show, choose this button to navigate the show's seasons and the episodes within each season. In this case, navigate to the episode you want to view, and then choose that episode's Watch Now button.

>> **Add to Watchlist:** For Amazon content, choose this button to add the content to your Amazon watchlist for later viewing.

FIGURE 6-10:
The details screen for a TV show.

For non-Amazon content, choosing the movie or TV show opens the app of the streaming service that offers the content, where you usually see at least a Play button.

Controlling playback

Use your Alexa Voice Remote (or the remote feature in the Fire TV mobile app) to control playback using the following buttons:

» **Play/Pause:** Pause and restart the content.

» **Rewind:** Go back in the content as follows:

- Press Rewind once to rewind by ten seconds.

- Press and hold Rewind for a few seconds, and then release the button to continuously rewind at low speed.

- While rewinding at low speed, press Rewind again to continuously rewind at medium speed.

- While rewinding at medium speed, press Rewind again to continuously rewind at high speed.

  While rewinding at high speed, press Rewind again to switch to the low-speed rewind.

- To stop rewinding, press Play/Pause.

» **Fast Forward:** Go forward in the content as follows:

- Press Fast Forward once to fast-forward by ten seconds.

- Press and hold Fast Forward for a few seconds, and then release the button to continuously fast-forward at low speed.

- While fast-forwarding at low speed, press Fast Forward again to continuously fast-forward at medium speed.

- While fast-forwarding at medium speed, press Fast Forward again to continuously fast-forward at high speed.

  While fast-forwarding at high speed, press Fast Forward again to switch to the low-speed fast-forward.

- To stop fast-forwarding, press Play/Pause.

» **Volume Up:** Raise the volume one notch; press and hold Volume Up to raise the volume quickly.

» **Volume Down:** Decrease the volume by one value; press and hold Volume Down to decrease the volume quickly.

» **Mute:** Toggle the volume off and on.

If you're more of a voice person, you can control playback using the following commands:

» "Pause"

» "Resume"

>> "Rewind" (rewinds ten seconds)

>> "Fast forward" (fast-forwards ten seconds)

>> "Volume up"

>> "Volume down"

>> "Set volume to *X*" (replace *X* with the volume setting you want)

>> "Mute"

>> "Unmute"

Setting playback options

While you're watching a TV show or movie, you can press the Menu button to see a list of options for the content, as shown in Figure 6-11. Depending on the content, you see some or all of the following:

>> **Next Up:** For a TV show, play the next episode.

>> **Watch from Beginning:** Restart the current TV show or movie.

>> **Subtitles:** Toggle subtitles for the content (and choose a subtitle language).

>> **Audio:** Set audio options such as the language and the type of audio output (usually either Dolby Digital Plus or Stereo).

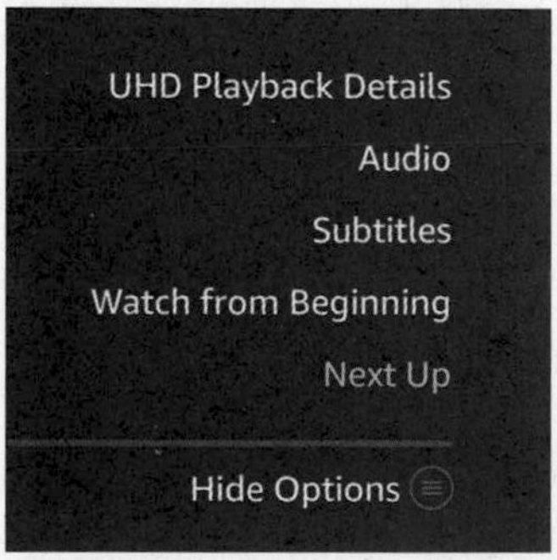

UHD Playback Details

Audio

Subtitles

Watch from Beginning

Next Up

Hide Options

FIGURE 6-11:
During playback,
press Menu to
see a list of
options.

Peeking at cast or music info

It's happened to all of us a million times: You're watching a TV show or movie and you think, "Hey, what's that actor's name?" or "Ooh, I like this music — I wonder who does it?" Normally, you either have to wait until the end of the show to get your answer in the credits (if you're lucky), or you pull out your smartphone and start Googling the actors or Shazaming the music.

Fortunately, Fire TV offers a better method. It's called X-Ray, and it enables you to take a quick peek at info related to the current scene (X-Ray Quick View) or the overall show (Full Screen X-Ray). X-Ray is powered by the famous (and famously comprehensive) Internet Movie Database (IMDb).

First, note that not all shows support X-Ray, although a remarkable number do. To know whether X-Ray is available, check out the show details. If you see the X-Ray label, as shown in Figure 6-12, then you're good to go.

FIGURE 6-12:
Look for the
X-Ray label in the
show's info.

The Marvelous Mrs. Maisel

IMDb 8.8/10

Pilot In 1958 New York, Midge Maisel's life is on track- husband, kids, and elegant Yom Kippur dinners in their Upper West Side ...

56 min   Air Date: March 16, 2017   16+   X-Ray   CC   UHD

X-Ray label

When you're watching an X-Ray-able show and you just need to know more about what's going on, display X-Ray Quick View by pressing the navigation ring's Up button on your Alexa Voice Remote. Fire TV continues to play the show, but along the bottom of the screen you see the current scene's cast members and a tile for any music that plays in the current scene. Figure 6-13 shows an example. To hide the X-Ray info, press the Down button.

FIGURE 6-13:
Press the
navigation ring's
Up button to see
the current
scene's cast and
music.

Rachel Brosnahan
Miriam 'Midge' Mai...

Tony Shalhoub
Abe Weissman

Marin Hinkle
Rose Weissman

Teach Me Tonight
The McGuire Sisters

01:00/55:51   **Ultra** HD **HDR**   View All X-Ray   Options

TIP

To display X-Ray Quick View and pause the show, press the Play/Pause button instead.

If you're *really* curious about the show, display Full Screen X-Ray by pressing the Up button twice (or by pressing Play/Pause and then pressing Up). Fire TV pauses the show and displays Full Screen X-Ray, as shown in Figure 6-14. There are four tabs you can traverse here:

>> **Scenes:** Displays a list of the show's scenes, which you can use to navigate quickly from one scene to another.

>> **In Scene:** Shows the cast members and music that are part of the scene you're currently watching.

>> **Cast:** Shows a complete list of the show's cast.

>> **Music:** Shows a complete list of the songs played during the show.

FIGURE 6-14:
An example of Full Screen X-Ray.

When you're done, you can return to the show by pressing the remote's Play/Pause button.

Viewing mobile content on your Fire TV Stick device

If you've got TV shows or movies on your mobile device, wouldn't it be great to watch that content on your big-screen TV? It sure would, and it just may be possible. Fire TV supports a feature called *mirroring* that takes whatever is displayed on a mobile device screen and also displays it on the Fire TV device. Sounds extremely cool, doesn't it? But there are some catches:

>> Mirroring only works with Fire TV Stick. If you have a Fire TV Cube or a Fire TV Edition device, you're out of luck.

>> Your mobile device must be an Android phone (running Android 4.2 or later), a Fire HD tablet, or a Fire phone. Sorry, iOS isn't supported.

To get started, first put your Fire TV Stick device into Display Mirroring mode. You have two choices:

>> Choose Settings ➪ Display & Sound ➪ Enable Display Mirroring.

>> Press and hold the Home button to display the Fire TV shortcuts, and then choose the Mirroring icon, as shown in Figure 6-15.

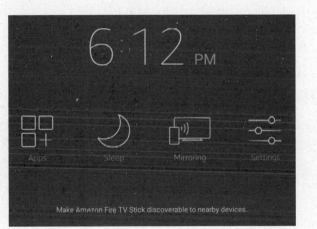

FIGURE 6-15:
Press and hold Home, and then choose Mirroring.

Now turn to your mobile device, which should be within about 30 feet of your Fire TV Stick, registered to the same Amazon account, and connected to the same Wi-Fi network. Open the mobile device's settings and activate the device's screen mirroring option. The name of the option varies depending on the version of Android your device uses, but here are some possibilities to search for:

>> Cast Screen

>> Display Mirroring

» Miracast

» Quick Connect

» Screen Mirroring

» Screen Sharing

» Smart View

» Wireless Display

» Wireless Mirroring

REMEMBER

If you can't find any settings with one of these names, it means your mobile device doesn't support screen mirroring. Note, in particular, that you won't find a screen mirroring setting in any recent version of the Fire tablet, because Amazon has removed that feature.

When you're done mirroring the mobile device screen to your Fire TV Stick, press any button on the Alexa Voice Remote to take your Fire TV Stick out of Display Mirroring mode.

Connecting Fire TV to a second screen device

In modern streaming media lingo, a *second screen experience* means you watch the media on a main screen (such as your TV), and you use a second screen (such as tablet or smartphone) to control playback and display extra info about the media, such as the cast and music info I talk about earlier (see "Peeking at Cast or Music Info").

The second screen experience is what all the cool kids are doing nowadays, so if you want to be cool, too, then you need the following:

» A second screen device, which can be a Fire HD tablet (version 6 or later); a Kindle Fire HD (second generation or later) or HDX; or a Fire phone.

» The second screen device and your Fire TV device connected to the same Wi-Fi network. (Actually, this is optional, but everything seems to work better if both devices are on the same network.)

» The second screen device and your Fire TV registered to the same Amazon account.

If you've checked off all those boxes, then your first step is to configure your Fire TV device to use a second screen device. Choose Settings ⇨ Display & Sounds, and then turn the Second Screen Notifications setting to On.

Now pick up your second screen device, use Amazon Prime Video to open the TV show or movie you want to watch, and then tap the Watch on Fire TV button.

The content begins playing on your Fire TV, while your second screen device displays playback controls and X-Ray info (if available) about the content.

REMEMBER

On your second screen device, you're free to switch to a different app at any time. The TV show or movie will continue to play on your Fire TV.

IN THIS CHAPTER

» Getting to know Fire TV Recast

» Setting up Fire TV Recast in your home

» Watching live over-the-air TV shows

» Recording and viewing over-the-air TV shows

» Managing your Fire TV Recast

Chapter **7**

Watching and Recording Shows with Fire TV Recast

D oes anyone watch live broadcast TV anymore? Oh, sure, lots of people still tune in to sporting events, election coverage, and other major events as they happen, but it's becoming increasingly rare to watch primetime shows and other standard TV fare when they first air. Instead, like the good digital citizens we've all become, we record what we want to watch and then play it back when it suits us and our lifestyles.

In this chapter, you discover the Amazon device that lets you watch and record live TV shows without requiring an expensive cable package. It's called Fire TV Recast, and its job is to view and record shows that you pick up using an HDTV antenna. It's the perfect complement to your existing Fire TV system, and you explore how it works in this chapter.

What Is Fire TV Recast?

In Chapter 10, I take you through a ten-step program for *cutting the cord* (ditching your cable TV account). If that sounds drastic, remember that you'll have access to free services such as Amazon Prime Video (if you're an Amazon Prime member), IMDb TV, and YouTube; subscription services such as Hulu and Netflix; and whatever over-the-air channels your HDTV antenna can bring in. Oh, and you also have that extra money in the bank from not having to pay an exorbitant cable bill each month. Sweet!

REMEMBER

Over-the-air (often shortened to *OTA*) refers to a signal that is broadcast wirelessly. So over-the-air TV refers to a TV signal broadcast from a station and picked up by an antenna.

Ah, but there's one little thing missing from this otherwise idyllic scenario. Streaming services enable you to watch content anywhere, anytime, but that's not true of over-the-air programming, which only airs once (usually) at a set time (what too-full-of-themselves network execs used to call "appointment television"). If you want the freedom and convenience of watching over-the-air shows whenever you like and however often you like (and I know you do), then you need to add another bit of equipment to your setup: a digital video recorder (DVR).

A DVR is a device that accepts an incoming video feed (such as the signal from an HDTV antenna) and records a specified show to an internal or external hard drive. You can then play back the show when it's convenient for you and replay the show as often as you can stand it.

There are lots of DVR choices out there, but you really only need one that records over-the-air TV. Again, you'll find a bunch of over-the-air recorders on the market, but because you're already comfortably ensconced in the Fire TV market, it makes sense to go with Amazon's Fire TV Recast, which (as you can tell by the name) is designed to work seamlessly with Fire TV. For this to work, you need a Fire TV Stick, Fire TV Stick 4K, Fire TV Cube, or a Fire TV Edition device (such as a Fire TV Edition Smart TV or Fire TV Edition Soundbar).

REMEMBER

In a pinch, if you don't have a Fire TV device, you can also view your Fire TV Recast programs and recordings on an Amazon Echo Show or Amazon Echo Show 5. Just make sure that the Fire TV Recast and the Echo Show are on the same Wi-Fi network and are registered to the same Amazon account.

Fire TV Recast comes in two flavors:

>> **Two tuners, 500GB, 75 hours:** This configuration comes with two channel tuners, which means you can record up to two shows simultaneously. This Fire TV Recast has a 500GB internal hard drive, which means you can store up to about 75 hours of high-definition (HD) programs.

>> **Four tuners, 1TB, 150 hours:** This configuration comes with four channel tuners, which means you can record up to four shows simultaneously. This Fire TV Recast has a 1TB internal hard drive, which means you can store up to about 150 hours of HD programs.

REMEMBER

A Fire TV Recast can only record shows from an over-the-air connection. You can't use a Fire TV Recast device to record shows from your streaming services or other Fire TV content.

Getting Ready for Fire TV Recast

Before getting started with the Fire TV Recast setup, here's a list of what you need to make the Fire TV Recast thing happen:

>> A Fire TV Recast device (but you knew that already).

>> An HDTV antenna (indoor or outdoor).

>> A coaxial cable.

>> A Wi-Fi router that's connected to the Internet.

>> A Fire TV Stick, Fire TV Stick 4K, Fire TV Cube, Fire TV Edition Smart TV, Fire TV Edition Soundbar, Echo Show, or Echo Show 5; make sure the device is connected to your Wi-Fi network.

>> An account with Amazon.

>> A mobile device (such as an iPhone, iPad, Android phone or tablet, or an Amazon Fire Tablet) running the Fire TV mobile app and connected to your Wi-Fi network.

It's worth mentioning here that technically you don't need a Fire TV device to use Fire TV Recast. Instead, you can watch and record over-the-air programs using only the Fire TV mobile app (or an Echo Show or Echo Show 5, if you have one). However, you *do* need a Fire TV device to manage the Fire TV Recast recordings, so it's worth including Fire TV in the loop.

Positioning Fire TV Recast

One of the things that often confuses folks who are new to Fire TV Recast is that the device doesn't have an interface of its own that you can use to set up the device or record shows. Many people expect that they can connect Fire TV Recast to a TV and then control the device from there. Nope, sorry. Instead, the Fire TV Recast uses your Wi-Fi network to connect to your Fire TV device and your Fire TV mobile app, and you watch and record over-the-air shows and customize Fire TV Recast through either the Fire TV interface or the Fire TV mobile app.

In other words, your Fire TV Recast is a set-it-and-forget-it device that, once configured, you can tuck away and never think about again. Okay, I hear you ask, but tuck away *where*, exactly? To answer that question, you need to know that your Fire TV Recast requires two physical connections, as shown in Figure 7-1:

>> A coaxial cable connection to an HDTV antenna

>> A power cable connection to an AC outlet

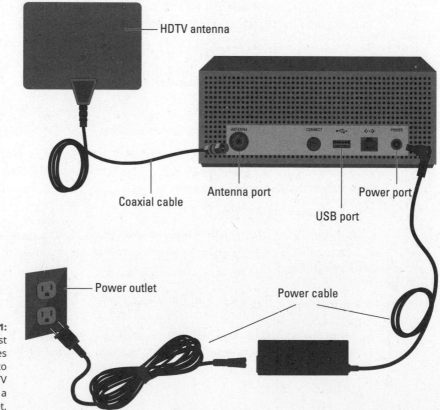

FIGURE 7-1:
Fire TV Recast
requires
connections to
both an HDTV
antenna and a
power outlet.

106   PART 2 **Watching Fire TV**

The HDTV antenna and the Fire TV Recast cable connections are what determine where you position your hardware in your home:

>> For best results with an indoor antenna, mount the antenna on or near a window.

>> The Fire TV Recast streams its recorded shows over your Wi-Fi network, so you can position it anywhere you like that's within the range of your Wi-Fi router (although, not surprisingly, the stronger the Wi-Fi signal, the better).

>> Make sure the Fire TV Recast is close enough to your HDTV antenna for the antenna's coaxial cable to reach.

>> Make sure the Fire TV Recast is close enough to a power outlet for the device's power cord to reach.

Setting Up Fire TV Recast

Run through these steps to get the Fire TV Recast device connected and ready to record your over-the-air programs:

1. **On your mobile device, open the Fire TV mobile app.**

2. **Tap Set Up a Fire TV Recast (U.S. Only).**

    As that "U.S. Only" parenthetical implies, as of this writing the Fire TV Recast can only be used with Amazon accounts based in the United States. Amazon has not announced any plans to bring Fire TV Recast to other countries.

    The app now runs through a few introductory screens, so tap Next until your mobile device asks for permission to allow Fire TV Recast to access your location.

3. **Tap Continue and then tap Allow (Android) or OK (iOS).**

    The Fire TV mobile app displays the Antenna Range screen.

4. **Tap the range of your antenna if you know it (if you don't know the range of your antenna, skip it), and then tap Next.**

    The Fire TV mobile app takes you through a series of screens that offer advice on choosing a location for your Fire TV Recast and your HDTV antenna.

5. **Run a coaxial cable from your HDTV antenna to the Antenna port on the back of the Fire TV Recast (refer to Figure 7-1), and then tap Next.**

6. **Connect the Fire TV Recast power cord to the power outlet, and then tap Next.**

The Fire TV mobile app now tells you to wait until the LED in front of the Fire TV Recast pulses white (see Figure 7-2) to indicate setup is in progress. Note that this may take as long as a couple of minutes, so put on your Patient hat for this step.

FIGURE 7-2:
When Fire TV Recast is connecting to your HDTV antenna, the front LED pulses white.

LED

7. **Tap the link that says The Light Is Pulsing White.**

The Fire TV mobile app connects to the Fire TV Recast and adds the device to your network. The app also registers your Fire TV Recast on your Amazon account and checks for updates to the Fire TV Recast software. If you see the message "Updating Fire TV Recast," put your Patient hat back on and wait until the update is complete.

WARNING

While you see the "Updating Fire TV Recast" message, do *not* unplug the Fire TV Recast from the power outlet or you may render the device unusable.

When the update (assuming there is one) is finished, Fire TV Recast starts scanning your antenna's over-the-air signal for available TV channels, and you see the progress of the scan in the Fire TV mobile app. When the scan is complete, the Fire TV mobile app opens the Channel Scan screen, which displays a list of the over-the-air channels it found.

8. **If the number of channels seems low, try moving the HDTV antenna to a different location, and then tap Rescan in the Channel Scan screen.**

Feel free to repeat Step 8 as often as you want. If you're disappointed by how many channels you get, remember that over-the-air signals are limited by both your geography (that is, how far away you are from the station transmitters) and topography (there may be local structures that are getting in the way of one or more signals).

9. **When you're done with the channel scan, tap Next.**

The All Done screen appears.

10. **Tap Done.**

REMEMBER

Your Fire TV Recast should pair automatically with your Fire TV or Fire TV Edition device (or your Echo Show). If the pairing doesn't work, for some reason, make sure the Fire TV Recast and Fire TV device are on the same network and registered to the same Amazon account, and then use the Fire TV mobile app to tap Settings ⇨ Live TV ⇨ Live TV Sources ⇨ Fire TV Recast ⇨ Pair a Fire TV Recast.

Watching and Recording Over-the-Air TV

With your Fire TV Recast connected to your antenna and paired with your Fire TV device and Fire TV mobile app, you're ready to start watching and recording over-the-air TV shows. Note that although most people control these Fire TV Recast functions through the Fire TV interface, you can also watch and record shows using the Fire TV mobile app or an Echo Show or Echo Show 5.

Managing your over-the-air channels

Before you access your over-the-air channels, there are a few channel-management chores you should run through to optimize your viewing and recording experience. In the next three sections I talk about checking channel signal strength, designating favorite channels, and hiding channels.

Checking channel signal strength

In the "Setting Up Fire TV Recast" section, earlier in this chapter, I explain that you can reposition your HDTV antenna and then ask Fire TV Recast to rescan for available channels. That's a great way to ensure you're receiving the most signals that are available to you, but it's unfortunate that the Fire TV Recast doesn't tell you anything about the signal strength of the channels it locates.

The next-best thing is to check each channel's signal strength yourself. You don't need to check every channel; instead, just check the signal for each channel that you'll be watching or recording frequently.

Here are the steps to follow to check the signal strength of your most-used channels:

1. **On your Fire TV device, choose Settings ⇨ Live TV ⇨ Channel Management ⇨ Fire TV Recast.**

    Fire TV displays a complete list of your channels.

2. **Select the channel you want to check.**

   On the right side of the screen, Fire TV displays a preview of the channel and the Signal Strength value: good, poor, or unavailable.

3. **If the channel preview is bad quality and/or the Signal Strength value is poor, try repositioning the HDTV antenna until you get a good signal.**

4. **Repeat steps 2 and 3 for the other channels you want to check.**

Adding a channel to your favorites

As I discuss real soon now (see "Recording over-the-air shows," later in this chapter), Fire TV Recast extends your normal Fire TV interface with a channel guide that lists your available over-the-air channels. That's an easy list to navigate if your antenna only brings in a few channels, but the list quickly gets unwieldy when your channels number in the dozens.

One quick way to tame a wild channel guide is to mark one or more channels as favorites, which then handily appear at the top of the channel guide (in a section labelled Favorite Channels; see Figure 7-3) for easy access.

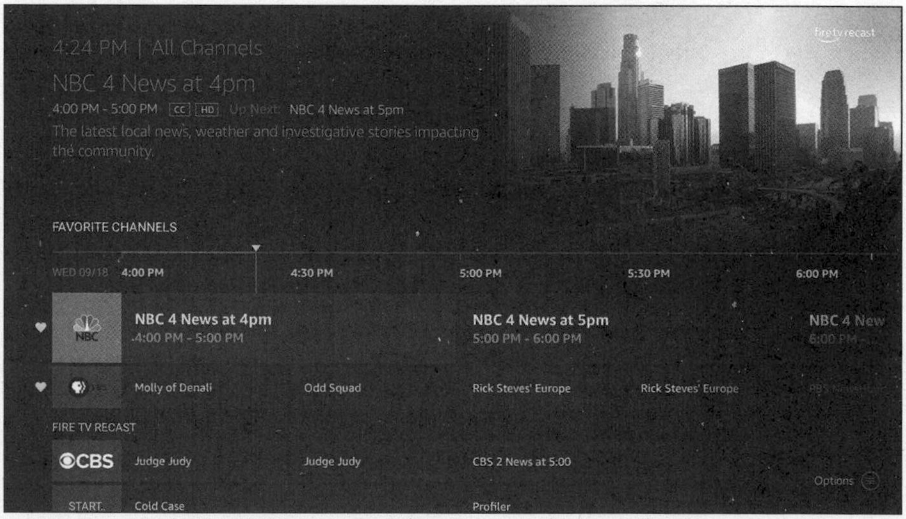

FIGURE 7-3:
Over-the-air channels marked as favorites appear in the Favorite Channels list.

Here are the steps to follow to mark one or more channels as favorites:

1. **On your Fire TV device, choose Settings ⇨ Live TV ⇨ Channel Management ⇨ Fire TV Recast.**

   Fire TV displays your over-the-air channels.

2. **Select the channel you want to mark as a favorite.**

3. **Select Add to Favorites.**

   Fire TV adds the channel to the favorites list.

   After you select the Add to Favorites command, the command name changes to Remove from Favorites, so you can use this command to toggle a channel on and off the list of favorites.

TIP

   You can also toggle a channel to and from the favorites list by pressing the Menu button on your Fire TV remote.

4. **Repeat steps 2 and 3 for the other channels you want to add to your favorites.**

Hiding a channel

If your HDTV antenna picks up quite a few stations, but it includes one or more stations that you never watch, you can make the channel guide easier and quicker to navigate by hiding those unwatched stations. Here's how it's done:

1. **On your Fire TV device, choose Settings ⇨ Live TV ⇨ Channel Management ⇨ Fire TV Recast.**

   Fire TV displays your channels.

2. **Select the channel you want to hide.**

3. **Select Hide Channel.**

   Fire TV removes the channel from the channel guide.

   After you select the Hide Channel command, the command name changes to Show Channel, so you can also use this command to reinstate a previously hidden channel.

TIP

   You can also hide or show a channel by pressing the Play/Pause button on your Fire TV remote.

4. **Repeat steps 2 and 3 for the other channels you want to hide.**

Watching over-the-air shows live

You probably picked up a Fire TV Recast because of its digital video recording capabilities, but the device is also a TV tuner, which means you can use your Fire TV Recast to watch live over-the-air shows on whatever channels your antenna picks up. Watching live TV is fine, but with your Fire TV Recast, you can also perform the following tricks:

›› Pause the show you're watching.

›› Rewind the current show all the way back to the point at which you started watching the show (up to 90 minutes).

›› Fast-forward the current show all the way up to the point at which the show is currently airing.

REMEMBER

You can watch live over-the-air TV on up to two devices (either Fire TV devices or Echo Show devices). This limit applies even if you have a Fire TV Recast that comes with four tuners.

Watching over-the-air shows using Fire TV

After your Fire TV Recast has paired with your Fire TV Stick, Fire TV Cube, or Fire TV Edition Smart TV or Fire TV Edition Soundbar, you'll notice that the menu bar at the top of the main Fire TV screen now comes with a DVR tab sandwiched between the Your Videos and Movies tabs (see Figure 7-4).

FIGURE 7-4:
Connecting a Fire TV Recast to your system adds the DVR tab to the Fire TV menu bar.

Use the Fire TV remote to select the DVR tab, which opens the DVR screen, shown in Figure 7-5. The On Now (Fire TV Recast) row shows thumbnail views of the current over-the-air TV shows. Use the Fire TV remote's navigation ring to navigate the shows in the On Now list. For each show you land on, Fire TV displays the name, start and stop times, and a brief description. When you land on a show you want to watch, press the Select button. To return to the On Now list, press the Back button on the Fire TV remote. If you prefer to go directly to the channel guide (which I discuss below), press the Fire TV remote's Menu button and then select Channel Guide in the menu that appears.

FIGURE 7-5: Selecting the DVR tab opens the DVR screen.

The Home screen's On Now list may also include other live TV streams that are included in your subscribed services. Alas, although you can use On Now to select one of these streams to watch, you can't record a stream using Fire TV Recast.

Alternatively, you can select the Live tab in the Fire TV Home menu bar. The Live screen that appears contains not only your Fire TV Recast over-the-air channels, but also any other live content from your other Fire TV services.

When this book was written, the Live command was a new feature that had not yet been rolled out to all Fire TV customers. If you don't see the Live command on your Fire TV menu bar, it means this feature isn't available in your area just yet.

When you navigate to a show using either the On Now row or the channel guide, you can quickly add the channel to your favorites list by pressing the Fire TV remote's Menu button and then selecting the Add to Favorite Channels command.

Watching over-the-air shows using the Fire TV mobile app

If you don't have a Fire TV device or your Fire TV device is up two floors and you're feeling lazy, no problem: You can watch your Fire TV Recast over-the-air shows right from the comfort of the Fire TV mobile app. Here's how it's done:

1. **Open the Fire TV mobile app on your mobile device.**

2. **Tap On Now.**

   The Fire TV mobile app displays the On Now screen, which shows thumbnail views and descriptions of the current over-the-air shows.

3. **Tap the show you want to watch.**

   The Fire TV mobile app starts playing the show.

Watching over-the-air shows with voice commands

With the Fire TV remote in hand, you can press the Voice button and then speak any of the following commands to watch over-the-air TV:

"Select DVR."

"Select Live."

"Show what's on now."

"Show the channel guide."

"Next" (to scroll to the next On Now or Channel Guide screen).

"Previous" (to scroll to the previous On Now or Channel Guide screen).

"Tune to [*signal*]" (where [*signal*] is the show's name or network).

"Pause."

"Play."

"Rewind *X* seconds" (where *X* is the number of seconds you want to go back).

"Fast-forward *X* seconds" (where *X* is the number of seconds you want to go forward).

Watching over-the-air shows using an Echo Show

If you have an Echo Show (or Echo Show 5) connected to your Fire TV Recast, say, "Alexa, show the channel guide" to bring up the On Now screen, which lists all your over-the-air channels. If you have more over-the-air channels than will fit on a single screen, say, "Alexa, next" to display the next screenful (and "Alexa, previous" to go back a screen). When you see the show you want to watch, say, "Alexa, tune to *signal*," where *signal* is the name of the show, the name of the show's network, or the show's number in the On Now list.

Setting recording options

Before I show you how to record over-the-air programs, you should take a minute or three to peruse and, if needed, adjust the Fire TV Recast options for recording TV series. For example, you can configure Fire TV Recast to stop all your series recordings a little after each show's scheduled stop time. Note that Fire TV Recast keeps track of two types of recording options:

>> **Default Recording Options:** Apply to every TV series you record (although they're overridden by the Series Recording Options that I discuss next); see Figure 7-6.

>> **Series Recording Options:** Apply only to the recording of a specific series.

FIGURE 7-6:
The Fire TV
Recast default
recording
options.

DEFAULT RECORDING OPTIONS

Start Recording
On Time (Default)

Stop Recording
On Time (Default)

Keep At Most
15

Recording Preference
All Episodes

Image Quality
HD Preferred

Live TV

Select when recordings will
start for a program.

Here are the available options:

>> **Start Recording:** Choose On Time to start the recording at the show's
scheduled start time, as shown in the channel guide (this is the default
setting). However, you may find that one or more shows mis-time the
beginning and start a minute or two earlier, which means you miss the
beginning. Cue the gnashing of teeth. In that case, use the Start Recording
option to select the number of minutes (1, 2, 3, 4, 5, 10, 15, 30, or 60) before
the show's scheduled start time that you want the recording to begin.

>> **Stop Recording:** Choose On Time to stop the recording at the show's
scheduled stop time as shown in the channel guide (this is the default setting).
If you find that one or more shows are running late (usually because a
previous show — I'm looking at *you* Sunday afternoon football games — ran
over its scheduled stop time), use the Stop Recording option to select the
number of minutes (1, 2, 3, 4, 5, 10, 15, 30, or 60) after the show's scheduled
stop time that you want the recording to end.

>> **Protect (Never Delete):** Choose No to enable series episodes to be deleted
(this is the default). If you want to ensure that a series' episodes can't be
deleted, choose Yes instead.

>> **Keep At Most:** Choose the maximum number of the most recent series
episodes you want Fire TV Recast to store (5, 15, 20, or All Available; the
default is 15). If you choose a number, then when you reach that maximum,
Fire TV Recast automatically begins deleting the oldest episodes. If you choose
All Available, Fire TV Recast keeps every episode in the series until it runs out
of storage space, at which point it begins deleting the oldest episodes to free
up room for new recordings.

If you find that the Keep At Most option is disabled, it means you've protected
the series by setting its Protect (Never Delete) option to Yes.

REMEMBER

» **Recording Preference:** Choose All Episodes to configure Fire TV Recast to record both first-run episodes and reruns (this is the default). If you don't want Fire TV Recast to record reruns, choose New Episodes instead.

» **Image Quality:** Choose the quality of the recordings: high definition (HD) or standard definition (SD). There are five choices:

- **HD Preferred:** Records in HD if it's available, but still records the show if it's presented in SD; this is the default.

- **HD Only:** Records only if the show is presented in HD; skips the recording if the show is presented in SD.

- **SD Preferred:** Records in SD if it's available, but still records the show if it's presented in HD.

- **SD Only:** Records only if the show is presented in SD; skips the recording if the show is presented in HD.

- **No Preference:** Records the show no matter what quality it's presented in.

Setting recording options using Fire TV

If you're using a Fire TV device, you can set your preferred recording options as follows:

» **Default recording options for all series:** Choose DVR ⇨ DVR Settings ⇨ Default Recording Options. (Alternatively, you can choose Settings ⇨ Live TV ⇨ Live TV Sources ⇨ Fire TV Recast ⇨ DVR ⇨ Default Recording Options.)

» **Recording options for a series:** Select DVR, navigate to and then select the series, and then select Series Recording Options. (Alternatively, highlight the series in the DVR screen, press the Menu button on the Fire TV remote, and then select Recording Options.)

Setting recording options using the Fire TV mobile app

You can also use the Fire TV mobile app to set your preferred recording options. However, the app doesn't let you set recording options for individual series. Instead, all you can do with the app is set the default recording options that apply to all series: Choose Settings ⇨ Fire TV Recast ⇨ Default Recording Options.

Recording over-the-air shows

Fire TV Recast makes it easy to watch live over-the-air TV, but it really shines when it comes to recording stuff to watch later when it's convenient for you. Fire TV Recast gives you many ways to record your favorite shows. In the next few sections, I take you through everything you need to know.

Watching and recording shows at the same time

One of the first questions folks new to Fire TV Recast ask is, "Can I watch live TV and record another show at the same time?" The answer is a resounding "Yes!" although how many shows you can watch and record simultaneously depends on your Fire TV Recast device.

If you have Fire TV Recast with two tuners, you have the following watch-and-record options:

>> Watch one live and one recorded program on different devices and record one program in the background.

>> Watch two recorded programs on different devices and record up to two programs in the background.

If you have a Fire TV Recast with four tuners, you can watch and record simultaneously as follows:

>> Watch one live and one recorded program on different devices and record up to three other programs in the background.

>> Watch two recorded programs on different devices and record up to four programs in the background.

>> Watch two live programs on different devices and record up to two other programs in the background.

Recording over-the-air shows using Fire TV

If you're using a Fire TV device, here are the various methods you can use to record a series or episode:

>> From the Live screen or On Now list, highlight a show, press the Fire TV remote's Menu button, select Record, then select either This Episode or Full Series.

>> From the channel guide, highlight a show that's currently on-air, press the Fire TV remote's Menu button, select Record, and then select either This Episode or Full Series.

>> From the channel guide, navigate to a show that's scheduled to air in the future and then select the show to open its info screen. Select either Record Series or Record Episode.

>> From the Fire TV Home screen, search for the show you want to record. (I explain how to search Fire TV in Chapter 6.) In the search results, use the Channels list to navigate to the show you want; then select the show. If the show is currently on air, press the Fire TV remote's Menu button and then select Record. Select either This Episode or Full Series.

How do you know whether a given program is recording now or is scheduled to be recorded? Fire TV gives you two indicators:

>> If a show is being recorded now, you see a solid red dot in the show's thumbnail in the On Now list or the Channel Guide.

>> If a show is scheduled to be recorded, you see a hollow red dot in the show's thumbnail in the On Now list or the Channel Guide.

TIP

If a recording is in progress and it looks like the program will extend past the scheduled stop time, stay calm, open the Fire TV DVR tab, and then navigate to the recording program. Press Menu on your Fire TV remote, select Recording Options, and then adjust the Stop Recording value to extend the recording up to one hour. Whew!

Recording over-the-air shows using the Fire TV mobile app

Unfortunately, the Fire TV mobile app offers only a limited number of options for recording programs. Specifically, you can't search for programs to record, nor can you access a channel guide to schedule a recording. All you can do through the app is record a currently on-air program as follows:

>> Tap On Now, use the On Now screen to displays the thumbnail view of the show you want to record, tap More Options, and then tap Record.

>> While watching a show, tap the screen and then tap Record.

Recording over-the-air shows with voice commands

To record an over-the-air show by voice, use the Fire TV remote to press the Voice button and then say the following:

"Record [*series*]."

Make sure that [*series*] is the full title of the series you want to record. In case you're wondering, yes this command tells Fire TV Recast to record all episodes of the series you specified and, no, there isn't a voice way to tell Fire TV Recast to record just a single episode in the series.

To cancel a recording, press the Voice button and then say

"Cancel recording [*series*]."

Again, [*series*] is the full title of the series you want to stop recording.

Watching recorded over-the-air shows

After you've recorded one or more shows using Fire TV Recast, you can watch those shows anytime you like using Fire TV, the Fire TV mobile app, or an Echo Show. Your Fire TV Recast also lets you control playback by pausing, rewinding, and fast-forwarding the show.

Watching recorded over-the-air shows using Fire TV

To watch a recorded show on your Fire TV device, follow these steps:

1. **Select the DVR tab in the Fire TV menu bar.**

   The DVR screen appears. This screen includes a list titled My Recordings, which contains all your Fire TV Recast recordings.

2. **Use the My Recordings list to highlight the recorded program you want to watch.**

3. **Press the Select button on the Fire TV remote.**

   Fire TV Recast begins playing the recorded show.

TIP

The Protect (Never Delete) option that I talk about earlier applies to all the episodes in a series, but what if you want to protect just a single episode? No sweat: Follow steps 1 and 2 to highlight the recording you want to preserve, press the Fire TV remote's Menu button, and then select Protect Recording.

Watching recorded over-the-air shows using the Fire TV mobile app

Whether you recorded an over-the-air show using a Fire TV device or the Fire TV mobile app, you can still watch any recording using the app, which is handy if you're not sitting by your TV. Here's what you do:

1. **Open the Fire TV mobile app on your mobile device.**

2. **Tap Recordings.**

   The Fire TV mobile app displays the Recordings screen, which shows thumbnail views and descriptions of the shows you've recorded.

3. **Tap the recording you want to watch.**

   The Fire TV mobile app starts playing the recording.

Watching recorded over-the-air shows using voice commands

To watch recordings via voice commands, pick up the Fire TV remote, press the Voice button, and then speak any of the following commands:

"Show my recordings."

"Next" (to scroll to the next My Recordings screen).

"Previous" (to scroll to the previous My Recordings screen).

"Play [*name*]" (where *name* is the show's name).

"Pause."

"Resume."

"Rewind X seconds" (where X is the number of seconds you want to go back).

"Fast-forward X seconds" (where X is the number of seconds you want to go forward).

Watching recorded over-the-air shows using an Echo Show

If you have an Echo Show (or Echo Show 5) connected to your Fire TV Recast, say, "Alexa, show my recordings" to bring up the My Recordings screen, which lists all your recorded shows. If you have more recordings than will fit on a single screen, say, "Alexa, next" to display the next screenful (and "Alexa, previous" to go back a screen). When you see the recording you want to watch, say, "Alexa, tune to [*signal*]," where [*signal*] is the name of the show, the name of the show's network, or the show's number in the On Now list.

Managing Your Fire TV Recast

As I mention earlier, the Fire TV Recast is blissfully noninteractive, so most of the time you can set it up and then basically forget that it even exists. However, the Fire TV Recast does come with a few management features that you should know about, and I spend the rest of this chapter telling you about them.

Checking DVR storage

Your Fire TV Recast stores its recordings on an internal hard drive that contains either 500GB or 1TB of digital real estate, depending on which model of Fire TV Recast your purchased. Whichever size drive you have, it's a good idea to keep an eye on how much storage the drive is currently using because if the drive gets full, Fire TV Recast will automatically start deleting the oldest show episodes to make room for new recordings.

You can check how much space your recordings are currently using with either a Fire TV device or the Fire TV mobile app:

>> **Fire TV device:** Choose Settings ⇨ Live TV ⇨ Live TV Sources ⇨ Fire TV Recast ⇨ DVR Storage (see Figure 7-7).

>> **Fire TV mobile app:** Choose Settings ⇨ Fire TV Recast ⇨ About ⇨ Internal Storage.

>> **Voice:** Press the Voice button and then say, "How full is my DVR?"

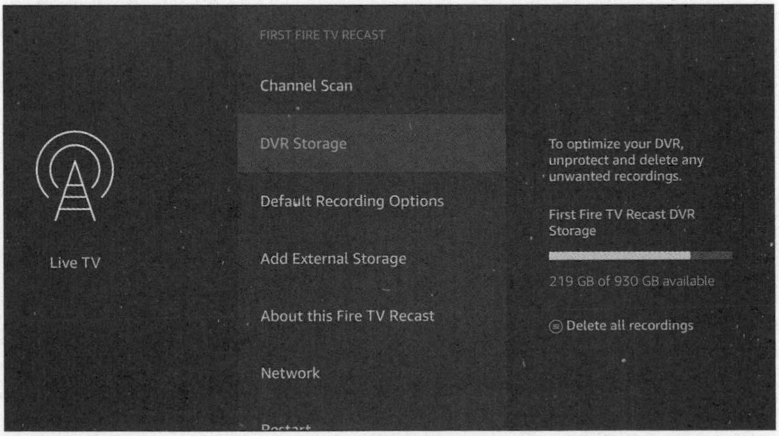

FIGURE 7-7:
Select the DVR Storage setting to check how much room you've used to store recordings.

FIRST FIRE TV RECAST

Channel Scan

DVR Storage

Default Recording Options

Add External Storage

About this Fire TV Recast

Network

Live TV

To optimize your DVR, unprotect and delete any unwanted recordings.

First Fire TV Recast DVR Storage

219 GB of 930 GB available

Delete all recordings

TECHNICAL STUFF

When you check your storage, don't be alarmed if the total storage is less than what you expect. For example, a Fire TV Recast with 1TB of storage may show something like 991GB as total storage. That "missing" storage is perfectly normal: It's a portion of the hard drive that Fire TV Recast used to install its internal software.

Adding external DVR storage

Whether you have a Fire TV Recast with 500GB or 1TB of internal storage, you may find yourself running out of room if you record and save a lot of shows. Hoarding, er, I mean *curating* a large collection of shows is very common, but don't let the size of your Fire TV Recast internal drive cramp your style. Instead, you can indulge your habit, sorry, *hobby* as needed by adding the largest external hard drive that your budget will allow. High-five!

Okay, so your first task is to get an external USB hard drive. Amazon says that it has "tested and currently recommends" just the following three drives:

>> Seagate Fast SSD 250GB

>> Toshiba Canvio for Desktop 3TB

>> WD 2TB Elements Portable External Hard Drive – USB 3.0

TIP

Amazon is just being cautious here, so don't feel you have to choose something from this list (as of this writing, the Toshiba has only limited availability anyway). For best performance without spending a fortune, stick with USB 3.0 drives.

Run through these steps on a Fire TV device to tell the Recast to use the new drive for storage:

1. Choose Settings ⇨ Live TV ⇨ Live TV Sources ⇨ Fire TV Recast ⇨ Add External Storage.

Fire TV prompts you to connect the external drive.

2. Connect the external hard drive to the USB port in the back of the Fire TV Recast.

Fire TV Recast detects the new drive automatically and then warns you that setting up the drive to use with Fire TV Recast will erase all the drive's content.

3. Select OK.

Fire TV Recast formats the drive and sets it up to use for storage.

REMEMBER

When Fire TV Recast starts using your external drive, it adds the drive's capacity to the capacity of the Fire TV Recast internal drive when it calculates total storage. Therefore, when you run the DVR Storage command (see the preceding section), the total available space you see is the sum of both the internal and external hard drives.

Checking scheduled recordings

I mention earlier that when you schedule a program for recording, Fire TV Recast lets you know by displaying the program in the channel guide with a hollow red circle. That's a nice touch, but what if you want to see *everything* that you've already scheduled to be recorded?

Easy money: Open the DVR screen, scroll down to the DVR Manager row, and then select Scheduled Recordings. Fire TV Recast displays the Scheduled Recordings screen with the Scheduled tab selected (see Figure 7-8 for an example). Select a show to see its details on the right. You can also select a show and then select Options (or press Menu on the remote) to tune to the station now, cancel the recording, or set recording options.

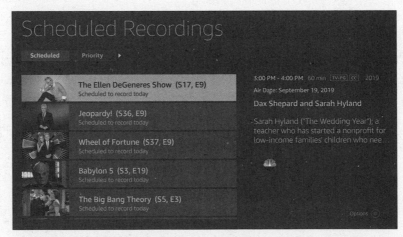

FIGURE 7-8: In the Scheduled Recordings screen, the Scheduled tab displays a list of the shows you've set up to record.

Changing recording priority

Depending on the number of tuners in your Fire TV Recast device, you can record up to two or four shows at the same time. Fair enough, but Fire TV Recast does *not* prevent you from scheduling more than your fair share of shows at a particular time. What happens if your Fire TV Recast supports a maximum of, say, two simultaneous recordings and you schedule *three* shows for recording at the same time? Fire TV Recast callously ignores one of the shows and records the other two.

How does Fire TV Recast decide which show to ignore? It uses the order in which you scheduled the recordings, with the first shows scheduled getting higher priority than the last shows you scheduled. What if one of those later-scheduled shows is something that you *really* want to record? Then you have to force Fire TV Recast to record it by manually changing the priority of your scheduled recordings. Here's how you do it:

1. **From the Fire TV Home screen, select DVR in the menu bar.**

2. **Scroll down to the DVR Manager row and select Recording Priority.**

   Alternatively, if you're already viewing your scheduled recordings as I describe in the preceding section, select the Priority tab.

   Fire TV Recast displays the Priority list, an example of which is shown in Figure 7-9.

3. **Highlight the show you want to change, and then press Select.**

4. **Use the Fire TV remote's navigation ring to move the show to the priority you want.**

   Press Up to move the show to a higher priority; press Down to move the show to a lower priority.

5. **When the show is in the priority position you want, press Select to set the new priority.**

6. **Repeat steps 3 through 5 to change the priority for any other shows you want to work with.**

FIGURE 7-9:
On the Scheduled Recordings screen, use the Priority tab to give one show priority over another when recording simultaneously.

Deleting recordings

When your Fire TV Recast hard drive becomes about 95 percent full, the device automatically starts deleting your oldest recordings whenever it needs to clear out some space for new programs. (This automatic deletion process also kicks in at 95 percent of your total storage when you've added an external drive, as I describe in the "Adding external DVR storage" section, earlier in this chapter.) Note that these automatic deletions do *not* apply to any recording that you've protected.

Having Fire TV Recast automate the deletion process takes one thing off your to-do list, but the cost of that convenience may be losing one or more treasured recordings that you forgot to protect. Instead, you can take matters into your own hands and delete no-longer-needed recordings yourself.

WARNING

Take extra care when deleting a recording because once when Fire TV Recast trashes the recording, there's no getting it back.

Here's how to delete recordings you don't need anymore:

>> **Fire TV:** In the DVR tab, use the My Recordings list to navigate to the program you want to delete, press the Fire TV remote's Menu button, and then select Delete.

>> **Fire TV mobile app:** Select the Recordings tab, scroll to the program you want to delete, tap the program's More Options button, and then tap Delete.

>> **Voice:** Press the Fire TV remote's Voice button and then say, "Delete [*name*]," where [*name*] is the full title of the program you want to delete.

Rescanning channels

If you move your existing antenna, get a new antenna, or move to a new home, you should tell Fire TV Recast to run a new channel scan to see what's available.

You can rescan channels using either a Fire TV device or the Fire TV mobile app:

>> **Fire TV device:** Choose Settings ➪ Live TV ➪ Live TV Sources ➪ Fire TV Recast ➪ Channel Scan.

>> **Fire TV mobile app:** Choose Settings ➪ Fire TV Recast ➪ Channel Management ➪ Channel Scan.

3
Going Beyond the Basics

IN THIS CHAPTER

» **Connecting Bluetooth and USB devices to Fire TV**

» **Adding parental control to make Fire TV more kid-safe**

» **Making Fire TV accessible**

» **Browsing the web with Fire TV**

» **Listening to music and viewing your vacation photos and videos**

Chapter **8**

Getting More Out of Fire TV

When you saw this book for the first time (on Amazon, right?), you may have asked yourself, "Does the world really need an entire book on Fire TV?" It's a good question, and if you thought at the time that Fire TV was only about streaming *Game of Thrones,* then you no doubt answered your own question with a snort, followed by "No, duh."

However, I hope this book shows you that Fire TV runs *way* deeper than merely streaming your favorite TV show. If so, then I've justified the time I spent writing this book. If not, then I'm certain this chapter will win you over. Here, you explore a long list of other features that you can do with Fire TV, including adding Bluetooth and USB devices, setting up parental controls, making Fire TV accessible, browsing the web right on your TV, listening to music, and even viewing your own photos and videos. It's a veritable Fire TV smorgasbord. Belly up!

Connecting Bluetooth Devices

Your Fire TV device supports a wireless technology called Bluetooth, which enables you to make wireless connections to other Bluetooth-friendly devices. In particular, Fire TV can connect to Bluetooth headphones, which enables you to watch a show without disturbing people nearby. You can also connect to a Bluetooth keyboard for easier text input.

In theory, connecting Bluetooth devices should be laughably easy: You bring the devices within 33 feet (10 meters) of each other (the maximum Bluetooth range), they exchange the necessary technical pleasantries, and then they connect without muss or fuss. In practice, however, there's usually at least a bit of muss (and sometimes some fuss, too), which usually takes one or both of the following forms:

>> **Making the devices discoverable:** Unlike Wi-Fi devices, which broadcast their signals constantly, most Bluetooth devices broadcast their availability — that is, they make themselves *discoverable* — only when you tell them to. This makes sense in many cases because you usually only want to connect a Bluetooth component, such as headphones, with a single device. By controlling when the Bluetooth gadget is discoverable, you ensure that it works only with the device you want it to.

>> **Pairing Fire TV and the device:** As a security precaution, Bluetooth devices need to be *paired* with another device before the connection is established. This is why making a device discoverable is also known as putting the device into *pairing mode.*

To make things even less convenient, Fire TV does have a few Bluetooth restrictions you should be aware of:

>> You can connect up to seven Bluetooth accessories to your Fire TV device. However, you can only pair one set of Bluetooth headphones at a time.

>> If you're using a Fire TV Stick, you can only pair with Bluetooth game controllers.

>> Fire TV isn't compatible with Bluetooth microphones.

>> Fire TV isn't compatible with Bluetooth Low Energy (BLE) devices.

>> Bluetooth accessories can only pair with one other device at a time, so if you've previously paired your Bluetooth device with something else, you have to disconnect that other device before you can pair it with your Fire TV device.

You may think you can use Bluetooth to pair your smartphone or tablet with Fire TV to watch, say, a video on your TV. Nope, sorry. Fire TV doesn't do that. However, it *is* possible to display mobile content on your Fire TV device. I show you how in Chapter 6.

In the following sections, I walk you through how to pair Bluetooth devices, connect to paired devices, and unpair Bluetooth devices.

Pairing your Bluetooth devices

Okay, so what *can* you pair with Fire TV? The four main categories of useful Bluetooth accessories for Fire TV are headphones, keyboards, remotes, and game controllers. The next four sections provide the pairing details.

Bluetooth headphones

If you want to watch a TV show or movie, or even listen to music (see "Listening to Music," later in this chapter), but you don't want the sound to disturb anyone within earshot, then pumping the audio through a pair of Bluetooth headphones is the way to go.

Follow these steps to pair your Fire TV device with Bluetooth headphones:

1. **On your Fire TV device, choose Settings ⇨ Remotes & Bluetooth Devices.**

2. **Choose Other Bluetooth Devices.**

3. **Choose Add Bluetooth Devices.**

4. **Turn on the switch or press the button that makes the headphones discoverable.**

   Wait until you see the headphones appear on the Add Bluetooth Devices screen, as shown in Figure 8-1 (the "MPOW H12" device, which is a pair of headphones).

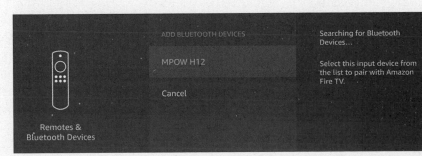

FIGURE 8-1: Wait until you see the headphones appear; then choose them to pair.

5. **Choose the Bluetooth headphones.**

Your Fire TV device pairs with the headphones.

When the pairing is complete, you see a Device Connected notification, and the headphones appear in the Other Bluetooth Devices list, as shown in Figure 8-2.

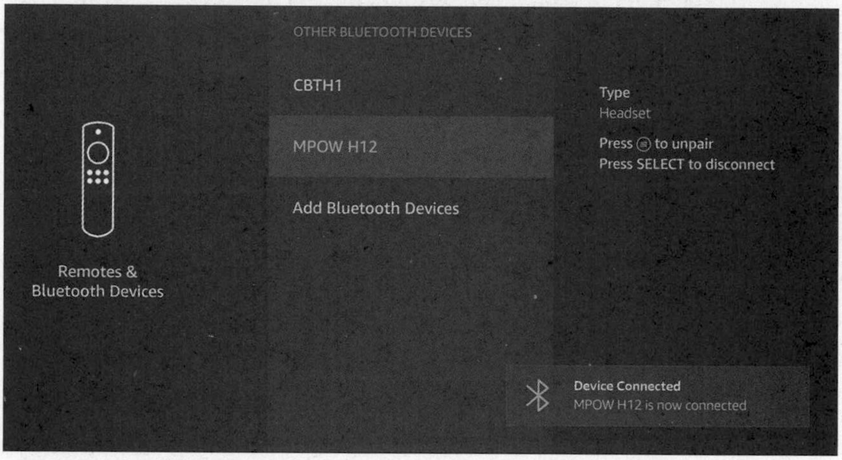

FIGURE 8-2:
The Other Bluetooth Devices screen maintains a list of your paired devices.

> **TIP**
>
> If you find that your Bluetooth headphones audio is out of sync with the video playback on your Fire TV, choose Settings ➪ Display & Sounds ➪ Audio Output ➪ Bluetooth Audio Sync. Press and hold the Alexa Voice Remote Voice button until Fire TV tells you the sync is complete.

Bluetooth keyboard

Using the Alexa Voice Remote navigation ring to peck out text one character at a time is no one's idea of a good time. Most people's preferred option is the keyboard feature in the Fire TV mobile app, but if you don't have your mobile device nearby, there's yet another way to easily enter significant chunks of text into the Fire TV interface: a Bluetooth keyboard. Here are the steps to follow to pair a Bluetooth keyboard:

1. **On your Fire TV device, choose Settings ➪ Remotes & Bluetooth Devices.**

2. **Choose Other Bluetooth Devices.**

3. **Choose Add Bluetooth Devices.**

4. **Turn on the switch or press the button that makes the keyboard discoverable.**

Wait until you see the keyboard show up in the Add Bluetooth Devices screen.

5. **Choose the Bluetooth keyboard.**

Fire TV displays a six-digit passkey, as shown in Figure 8-3.

FIGURE 8-3:
Fire TV displays a
passkey that you
type on the
Bluetooth
keyboard.

Bluetooth pairing request

To pair with:
Microsoft Bluetooth Mobile Keyboard 6000

Type on it:
748551, then press Return or Enter.

Cancel

6. **On the Bluetooth keyboard, type the passkey you see on your screen and press Return or Enter.**

Fire TV pairs with the keyboard and you see a Device Connected notification. The keyboard also appears in the Other Bluetooth Devices list.

REMEMBER

You can also pair a Bluetooth mouse with Fire TV. You can't use the mouse to navigate the Fire TV interface, but some apps are mouse-friendly (such as the Fire TV web browsers; see "Browsing the Web," later in this chapter).

A bonus you get with a Bluetooth keyboard is that you can also use the device to navigate the Fire TV interface. Here are the keys you can use:

Press	To
Up arrow	Move the selector up.
Right arrow	Move the selector to the right or fast-forward ten seconds during media playback; press and hold for continuous fast forward (same as the Fast Forward button on the remote).
Down arrow	Move the selector down.
Left arrow	Move the selector to the left or rewind ten seconds during media playback; press and hold for continuous rewind (same as the Rewind button).
Enter	Choose the selected item (same as the Select button).
Esc	Go back to the previous screen (same as the Back button).

Press	To
Spacebar	Toggle the current media between Play and Pause (same as the Play/Pause button).
Keyboard characters	Input text.

Bluetooth remote

If you lose your Alexa Voice Remote, you can purchase a new one from Amazon. There are also some third-party remotes that are compatible with Fire TV and even offer extra features, such as a built-in keyboard.

PAIRING A NEW ALEXA VOICE REMOTE

If you purchase another Alexa Voice Remote instead of a third-party remote, pair it with your Fire TV device by following these steps:

1. **On your Fire TV device, choose Settings ⇨ Remotes & Bluetooth Devices.**
2. **Choose Amazon Fire TV Remotes.**
3. **Choose Add New Remote.**
4. **Press and hold the new remote's Home button for ten seconds to put it into pairing mode.**

   Fire TV pairs with the new Alexa Voice Remote.

PAIRING A THIRD-PARTY REMOTE

Here are the steps to follow to pair your third-party remote with your Fire TV device:

WARNING

Note that only a few third-party remotes are compatible with Fire TV. Before buying a new remote, double-check that the device actually works with Fire TV.

1. **On your Fire TV device, choose Settings ⇨ Remotes & Bluetooth Devices.**
2. **Choose Other Bluetooth Devices.**
3. **Choose Add Bluetooth Devices.**
4. **On the remote, press the Home button for a few seconds to put the device into pairing mode.**

   After a short (hopefully!) delay, the remote appears in the Add Bluetooth Devices screen.

5. **Choose the remote.**

Fire TV pairs with the remote.

Bluetooth game controller

You can use a game controller not only to play games on Fire TV (no surprise there), but also to navigate the Fire TV interface.

WARNING

Before investing in a new game controller, make sure the device is compatible with Fire TV.

Here are the steps to follow to pair a game controller with your Fire TV device:

1. **On your Fire TV device, choose Settings ⇨ Remotes & Bluetooth Devices.**

2. **Choose Game Controllers.**

3. **Choose Add New Game Controller.**

4. **On the game controller, press the Home button for a few seconds to put the device into pairing mode.**

The game controller appears on the Add New Game Controller screen.

5. **Choose the game controller.**

Your Fire TV device pairs with the game controller.

Connecting to paired Bluetooth devices

If you've paired multiple Bluetooth devices of the same type (such as headphones), you can switch between them by connecting to the device you want to use. Similarly, if you've turned off a Bluetooth device, Fire TV automatically disconnects from the device, so to use the device again later you need to connect to it again.

Follow these steps to connect to a paired Bluetooth device:

1. **On your Fire TV device, choose Settings ⇨ Remotes & Bluetooth Devices ⇨ Other Bluetooth Devices.**

2. **In the Other Bluetooth Devices list, highlight the device you want to use.**

3. **Press Select.**

Fire TV connects to the device.

Unpairing a Bluetooth device

If you no longer want to use a paired Bluetooth device, you can unpair the device as follows:

1. **On your Fire TV device, choose Settings ⇨ Remotes & Bluetooth Devices ⇨ Other Bluetooth Devices.**

2. **In the Other Bluetooth Devices list, highlight the device you want to unpair.**

3. **Press Menu.**

   Fire TV asks you to confirm.

4. **Press Select.**

   Fire TV unpairs the device.

Connecting USB Devices

If your TV has one or more USB ports, you can connect some devices to those ports to accessorize your Fire TV system. What kinds of USB devices does Fire TV support? I'm glad you asked:

>> USB storage drive (Fire TV Cube and Fire TV Edition TVs only)

>> Wireless keyboard connected using a USB dongle

>> Wireless mouse connected using a USB dongle

>> Xbox 360 wireless gaming receiver

>> Wired gamepad (as long as it's not connected using a USB charger)

>> Flirc USB dongle (connects some third-party infrared [IR] remotes)

REMEMBER

A mouse? Yep, but it's not what you think because Fire TV doesn't support using a mouse to navigate the interface. However, some Fire TV apps and games are mouse-friendly, particularly the Fire TV web browsers (see "Browsing the Web," later in this chapter).

Adding a USB storage drive

A wireless USB keyboard may come in handy for text-heavy Fire TV operations, but the real star of the USB device show is USB storage. Your Fire TV device comes with its own internal storage, but if you augment that storage with an external USB hard drive or flash drive, then you get some significant advantages:

>> More storage for apps and games

>> The option to move some apps from internal to external storage

>> A longer live TV recording buffer, which gives you more options for pausing, rewinding, and fast-forwarding live TV. The larger the storage device you add, the longer the resulting live TV recording buffer.

WARNING

Fire TV will format the drive you use, which means that any data on the drive will be gone forever. Therefore, before proceeding, plug the drive into a computer and check that it doesn't contain any data you wouldn't want to lose. If the storage device does contain data you want to preserve, copy or move that data onto your PC.

Follow these steps to add a USB storage drive to your Fire TV Cube or Fire TV Edition TV:

1. **Insert the USB drive into a USB port on your Fire TV device.**

If required, be sure to plug the USB drive into a power outlet.

If Fire TV can't read the storage device, you see the screen shown in Figure 8-4.

> ### Unable to Read USB Device
>
> Your USB device is using an unsupported file system. TV can format the USB device to one of the two file systems. Note that all content on your USB device will be erased during formatting.
>
> Device Storage: Select this option if you intend to use apps stored on this USB drive. Once formatted, this USB device can only be used on this TV.
>
> External Storage: Select this option if you are storing content other than apps. The USB drive will be formatted to FAT32, enabling it to be used elsewhere.
>
> External Storage Device Storage Do Nothing

FIGURE 8-4: Fire TV displays this screen if it can't read the USB storage device.

2. **Choose an option to proceed.**

- **External Storage:** Choose this option to format the drive in a way that also enables you to add media, such as photos and videos, to the USB drive (see "Viewing Your Photos and Videos," later in this chapter).

- **Device Storage:** Choose this option to format the drive for exclusive use of Fire TV.

- **Do Nothing:** Choose this option to bail out of adding the USB drive.

If you choose either External Storage or Device Storage, Fire TV asks if you're sure you want to proceed.

3. **Choose Yes.**

   Fire TV formats the USB drive. When the operation is done, you see the Formatting Complete screen.

4. **Choose OK.**

   Fire TV begins using the USB drive for storage.

TIP

If you formatted the USB drive for one type of storage and later on you decide you'd prefer to use the other type of storage, you can switch right from Fire TV. Choose Settings, choose either Device & Software (for Fire TV Edition devices) or My Fire TV (for Fire TV Cube), and then choose USB Drive. Choose the Format to External/Internal Storage option.

Ejecting a USB drive

If you no longer want to use an external USB drive with your Fire TV device, don't just yank the drive cable out of its USB port. Instead, eject the drive safely by following these steps:

1. **Choose Settings.**

2. **Choose either Device & Software (for Fire TV Edition devices) or My Fire TV (for Fire TV Cube).**

3. **Choose USB Drive.**

4. **Choose Eject USB Drive.**

   Fire TV disconnects the drive and then displays the notification shown in Figure 8-5 when it's safe to disconnect the drive's USB cable.

FIGURE 8-5:
Fire TV displays this notification when it's safe to disconnect the USB drive.

USB Storage Ejected
You can now safely remove your USB drive from your TV.

Letting Your Kids Watch Fire TV

If your kids have access to your Fire TV device (or if they have Fire TV devices of their own), then you may be a bit worried about some of the content they may be exposed to in certain apps (such as YouTube). Similarly, you may not want your kids purchasing media or making in-app purchases in apps or games.

For all those and similar parental worries, you can sleep better at night by activating the parental controls on your Fire TV device. These controls restrict the content and activities that kids can see and do as follows:

>> Require a PIN (that you set up) for any purchases.

>> Require the PIN before launching apps.

>> Require the PIN before accessing Settings.

>> Restrict viewing of Amazon Video content based on ratings that you set.

Activating parental controls

Here are the steps to plow through to set up parental controls on your Fire TV device:

1. **Choose Settings ⇨ Preferences ⇨ Parental Controls.**

Fire TV displays the Parental Controls screen, which shows the Parental Controls setting as Off.

2. **Choose Parental Controls.**

Fire TV prompts you to enter a five-digit PIN, which Fire TV uses to prevent unauthorized access to apps and purchases.

WARNING

The PIN is an important safeguard for your children (and your sanity), so don't enter something easy to guess, such as 12345 or 00000.

TIP

The PIN interface, which is shown in Figure 8-6, arranges the numbers 0 through 4 on an image of the Alexa Voice Remote navigation ring. To enter a number, press the corresponding button on the navigation ring. For example, press Up to enter 1 and press Select to enter 0. To get the digits 5 through 9 displayed, press the Menu button.

FIGURE 8-6: The numbers indicate which buttons to press on the navigation ring to enter your PIN.

Enter New PIN

Press ⏯ to show PIN

1

4   0   2

3

⊗ Delete   ⏮ 5–9

This PIN will protect Amazon Video purchases on all registered devices. Visit amazon.com/pin to reset your PIN.

3. **Use the remote's navigation ring to enter the five digits of the PIN you want to use.**

    By default, Fire TV hides the PIN's digits with dots. To see the PIN's digits, make sure no youngster is looking over your shoulder and then press the Fast Forward button.

    TIP

    Fire TV prompts you to reenter the PIN.

4. **Repeat the PIN again to confirm.**

    Fire TV enables parental controls and displays a summary of the default protections that are now in place on your Fire TV device.

5. **Choose OK.**

Configuring parental controls

Fire TV enables parental controls with a default set of restrictions. To customize those restrictions, choose Settings ⇨ Preferences, enter your PIN, and then choose Parental Controls to display the Parental Controls screen. Figure 8-7 shows Parental Controls with the default settings.

Here are the settings you can play around with:

>> **Parental Controls:** Choose this setting and then enter your PIN to turn off parental controls.

>> **PIN-Protect Purchases:** When On, requires a PIN to make media, app, and game purchases.

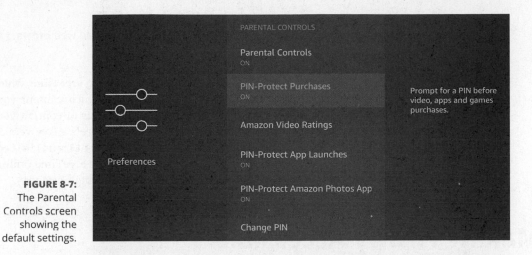

FIGURE 8-7:
The Parental
Controls screen
showing the
default settings.

>> **Amazon Video Ratings:** Enables you to specify which ratings applied to Amazon Video content your kids can view. All other ratings require a PIN to watch. Choosing this setting opens the Amazon Video Ratings screen, shown in Figure 8-8. For each rating, press Select to toggle between available (a checkmark icon) and restricted (a lock icon).

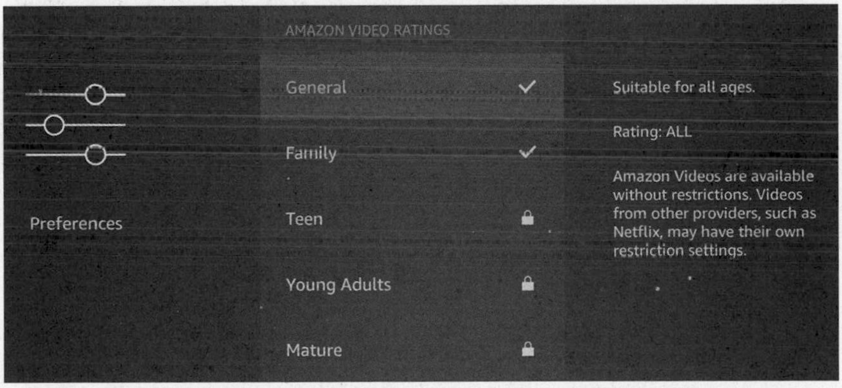

FIGURE 8-8:
Press Select on a
rating to toggle
kids' access to
content that
carries that
rating.

>> **PIN-Protect App Launches:** When On, requires a PIN to start apps.

>> **PIN-Protect Amazon Photos App:** Toggles PIN protection for Amazon Photos. If PIN-Protect App Launches is On, then PIN-Protect Amazon Photos App must also be On; if PIN-Protect App Launches is Off, then PIN-Protect Amazon Photos App can be either On or Off.

>> **Change PIN:** Choose this setting and enter your existing PIN to set a new PIN.

If you forget your PIN, you can reset it by directing your favorite web browser to www.amazon.com/pin.

For even more protection for your kids on Fire TV, set up Amazon FreeTime, which gives you a Parent Dashboard that you can use to monitor which content your children are accessing, use parental controls to restrict the types of content your kids can access, and set time restrictions on using Fire TV. There's a free version and a paid FreeTime Unlimited version (www.amazon.com/freetimeunlimited) that gives you extra controls and content (although as I write this FreeTime Unlimited is only available in the United States, the United Kingdom, and Germany).

Making Fire TV Accessible

For people with vision issues or mobility challenges, having Alexa as an integral part of Fire TV is a welcome feature because it means that much TV watching can be performed by voice only: Perfect vision or surgeon-steady hands are not required.

Similarly, Alexa also helps people with hearing problems, because Alexa often shows its results on the Fire TV screen.

Alexa's benefits to people with physical challenges are significant, but they don't mean that either Alexa or a Fire TV device is configured ideally for those same people. In the next couple of sections, I explore the accessibility features that can make Fire TV easier to use for people with hearing, mobility, and vision issues.

Enabling the Fire TV features for the vision impaired

If your eyesight isn't what it used to be, or if your eyesight problems go beyond relatively simple afflictions such as farsightedness or astigmatism, then you may need help making sense of what's happening on the Fire TV screen. Fortunately, a number of tools are available for enlarging screen items, making things easier to see, and even hearing audio transcriptions of what's on the screen.

Making Alexa more vision accessible

Both your Fire TV device and the Alexa app come with a few features that you can tweak or activate to work around vision problems:

» The Alexa app supports the vision accessibility features — such as dynamic type sizes and high contrast — available on your smartphone:

- **iOS:** Choose Settings ⇨ General ⇨ Accessibility and adjust the settings in the Vision section.

- **Android:** Choose Settings ⇨ Accessibility ⇨ Vision.

» You can operate the Alexa app using a screen reader:

- **iOS:** Choose Settings ⇨ General ⇨ Accessibility ⇨ VoiceOver and tap the VoiceOver switch to On.

- **Android:** Choose Settings ⇨ Accessibility ⇨ TalkBack and then tap the switch to On.

» You can configure the Fire TV Cube version of Alexa to make it more vision accessible. First, in the Alexa app, choose Devices ⇨ Echo & Alexa, tap your Fire TV Cube, and then tap Sounds. There are two techniques you can use:

- **Increase the volume for alarms, timers, and notifications.** Drag the Alarms, Timers, and Notification slider to the right to increase the default volume for these features.

- **Play a sound both when Alexa starts processing a request and when it finishes processing that request.** This gives you a better sense of when Alexa is processing your voice commands. To set this up, tap the Start of Request and End of Request switches to On, as shown in Figure 8-9.

Navigating Fire TV with VoiceView

You can use the VoiceView feature as a Fire TV screen reader. As you open each screen, VoiceView tells you the name of the screen, and you can move the selector to a screen element to have the element's name or text read out loud to you.

ACTIVATING VOICEVIEW

To activate VoiceView, follow these steps:

1. **Choose Settings ⇨ Accessibility.**

2. **Choose VoiceView Screen Reader.**

3. **Choose VoiceView.**

   Fire TV switches into VoiceView mode and a voice says, "VoiceView ready" and prompts you to press Play/Pause to see a tutorial for VoiceView. You can either press Play/Pause to run through the tutorial to learn the basics, or press Back and then press Select on the VoiceView switch once again.

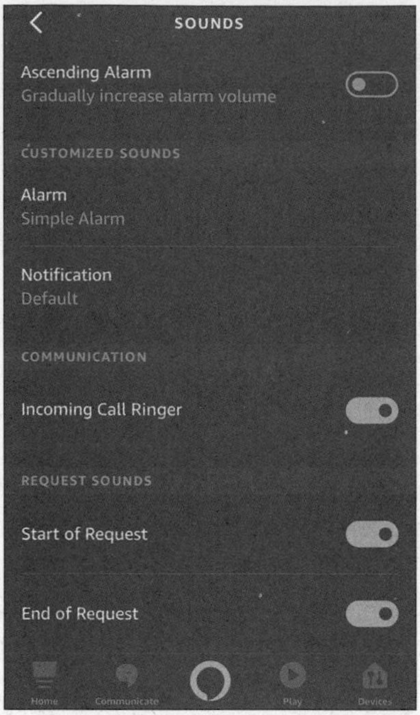

FIGURE 8-9:
Tap Start of
Request and
End of Request
to On to know
when Alexa is
processing your
voice commands.

NAVIGATING WITH VOICEVIEW

While VoiceView is active, how you interact with the screen doesn't change all that much. For example:

>> To hear the name or text associated with a screen element, use the navigation ring to move the selector to that element.

>> To open a screen element (for example, to launch an app), move the selector to the element and press Select.

You can also use the remote keys outlined in Table 8-1 to navigate your Fire TV device.

TABLE 8-1 ## Controlling VoiceView with Remote Keys

Press	To
Back + Menu for two seconds	Toggle VoiceView on or off.
Play/Pause	Mute VoiceView for the current element only (that is, you hear VoiceView again when you navigate to the next element).
Menu	Hear VoiceView tips or information related to the current screen.
Menu + Menu	Display options for the current screen (this is the same as pressing Menu once while VoiceView is turned off).

CONFIGURING VOICEVIEW SETTINGS

With VoiceView activated, the VoiceView Screen Reader screen now shows a long list of settings for customizing VoiceView to suit your style. Here's a quick summary of the settings:

>> **Reading Speed:** Sets the speed at which VoiceView reads the screen text.

>> **Verbosity:** Lets you specify which screen elements VoiceView mentions.

>> **Speech Volume:** Sets the VoiceView speech volume relative to the Fire TV current volume setting. By default, VoiceView uses 40 percent of the Fire TV current volume, but you can opt for a quieter playback (such as 20 percent of the Fire TV current volume), a louder playback (say, 70 percent), or the same volume as Fire TV (the Match Device Volume option).

>> **Sounds Volume:** Sets the VoiceView sound effects volume relative to the Fire TV current volume setting. This setting also defaults to 40 percent of the Fire TV current volume, but you can instead opt for a quieter or louder volume or a volume that matches Fire TV.

>> **Key Echo:** Specifies how VoiceView confirms what you've input when you're entering text using the onscreen keyboard. By default, VoiceView says each character as you type it (that is, it *echoes* the keys you press). You can also opt to have VoiceView echo only words (that is, VoiceView says each word as you complete it) or both characters and words.

>> **Punctuation Level:** Specifies whether and how much punctuation VoiceView includes in its descriptions.

>> **VoiceView Tutorial:** Runs a tutorial that gives you a quick lesson in how to use VoiceView.

WORKING IN VOICEVIEW'S REVIEW MODE

The navigation ring's Up, Down, Left, and Right buttons enable you to move through the screen, and as you land on each screen element, VoiceView speaks the element name or text. That's handy, but many screens include lots of "nonselectable" text, such as the summary info you see when you open a TV show or movie (plot summary, length, release year, and so on). To hear that text, you can switch VoiceView into Review mode by holding down your Fire TV remote's Menu button for two seconds. You hear VoiceView say "Review mode on" and VoiceView also changes how the navigation ring's buttons work:

>> **Left:** Navigates to the previous text element on the screen.

>> **Right:** Navigates to the next text element on the screen.

- **Up:** Increases the Review mode granularity.
- **Down:** Decreases the Review mode granularity.

"Granularity"? It's a word to warm the cockles of a nerd's heart, I know, but it just refers to the level of detail that Review mode uses when you navigate to some text on the screen. Here are the four levels of granularity in ascending order:

- **Character:** When you press Right or Left, VoiceView speaks the next or previous character of the text.
- **Word:** When you press Right or Left, VoiceView speaks the next or previous word of the text.
- **Sentence:** When you press Right or Left, VoiceView speaks the next or previous sentence of the text.
- **Element:** When you press Right or Left, VoiceView navigates to the next or previous element on the screen.

Zooming in with Screen Magnifier

You may find that although you can make out most of the items on your Fire TV device screen, the occasional icon or bit of text is just too small to decipher. You could always grab a nearby pair of binoculars to get a closer look at the section you can't make out, but Fire TV offers an electronic version of the same thing. It's called Screen Magnifier, and it enables you to zoom in on a portion of the screen for a closer look.

To activate Screen Magnifier, choose Settings ⇨ Accessibility ⇨ Screen Magnifier. Fire TV turns on Screen Magnifier and displays a list of remote key combinations that you can use to control Screen Magnifier. I summarize those key combos in Table 8-2.

TABLE 8-2

Screen Magnifier Remote Keys

Press	To
Back + Fast Forward	Toggle Screen Magnifier on or off
Menu + Play/Pause	Toggle zoom on and off
Menu + Fast Forward	Increase the zoom level
Menu + Rewind	Decrease the zoom level

Press	To
Menu + Up	Pan a zoomed screen up
Menu + Down	Pan a zoomed screen down
Menu + Left	Pan a zoomed screen left
Menu + Right	Pan a zoomed screen right

Switching to high-contrast text

If you have trouble reading text on the Fire TV screen, you may want to try a feature that was experimental as this book went to press: high-contrast text. With high-contrast text, Fire TV formats all regular screen text as follows:

>> If the screen background is dark, Fire TV uses only white text with a black border.

>> If the screen background is light, Fire TV uses only black text with a white border.

Note that when I say "regular" text, I'm talking about descriptive text, titles, settings, and so on. High-contrast text doesn't apply to logos, artwork, or any proprietary text.

To switch to high-contrast text, choose Settings ⇨ Accessibility and then choose High-Contrast Text to switch that setting to On.

"Hearing" the action with audio descriptions

If you have trouble seeing what's happening during a show, Fire TV offers a feature called Audio Description, which adds an audio track that describes what's happening on the screen. You can enable Audio Description for over-the-air channels or Amazon Prime Video content. Here are the steps to follow:

1. **Choose Settings ⇨ Accessibility.**

2. **Choose Audio Description.**

3. **To include the descriptive audio track on over-the-air TV, choose Over the Air Channels and then choose a volume level (Low, Medium, or High).**

   You also see a setting labeled Off, which is the one to choose if you later want to disable Audio Description for over-the-air TV.

4. **To include the descriptive audio track on Amazon Prime Video content, choose Prime Video to switch that setting to On.**

   Note that the Prime Video setting doesn't offer a volume level.

   Fire TV now displays a descriptive audio track on the source(s) you selected.

Checking out the Fire TV features for the hearing impaired

If your hearing has deteriorated over the years, or if you have a hearing impairment in one or both ears, hearing Alexa's responses and enjoying Fire TV videos can be a challenge. Fortunately, help is at hand. Fire TV has a few tools that you can configure to help or work around your hearing issues. The next few sections give you the lowdown on the hi-fi.

Making Alexa more hearing accessible

Fire TV and the Alexa app come with a few features that you can customize or turn on to work around hearing impairments:

>> The Fire TV Cube Light Ring provides a visual indicator of the status of your device.

>> The Fire TV Cube and most TVs have physical buttons you can press to adjust the volume.

>> You can configure the Fire TV Cube version of Alexa to increase the volume for alarms, timers, and notifications. In the Alexa app, choose Devices ➪ Echo & Alexa, tap your Fire TV Cube, tap Sounds, and then drag the Alarms, Timers, and Notification slider to the right to increase the default volume for these features.

>> You can configure Fire TV to display subtitles on videos (see "Enabling subtitles" later in this chapter).

>> You can connect Fire TV with Bluetooth headphones, as I describe earlier in this chapter (see "Pairing Bluetooth headphones").

FIRE TV WITH HEADPHONES

If you suffer from only minor hearing loss, you may be able to compensate by turning up the volume of your Fire TV device. However, this may be impractical if there are other people nearby and you don't want to disturb them, and it may be useless if your hearing loss is severe.

In such cases, an often better solution is to use headphones. Because the sound from the headphones takes a shorter and more direct path to your ear, it can make the sounds sharper and easier to discern — and it has the added advantage of not disturbing anyone within earshot.

If you decide to invest in some headphones, here are a few pointers to bear in mind:

- If you use an in-the-ear (ITE) hearing aid, look for *ear-pad headphones* (also called *on-ear headphones*), which rest on your ears.

- If you use a behind-the-ear (BTE) hearing aid, you'll need to move up to *full-size headphones* (also called *full-cup, ear-cup,* or *over-the-ear headphones*), which are large enough to cover not only your ears but also your hearing-aid microphones.

- With any type of hearing aid, you need to guard against *feedback,* where amplified sounds from your hearing aid "leak" out and bounce off the headphones back into the hearing aid. The cycle repeats until a painful feedback squeal is the result. To prevent feedback, get headphones that use foam ear pads, which reduce the chance of sound being reflected into the hearing aid.

- If your hearing aid comes with a *telecoil mode* (which enables the hearing aid to process sounds sent electromagnetically), be sure to get telecoil-compatible headphones (which broadcast sounds electromagnetically).

- Consider getting noise-canceling headphones, which virtually eliminate background noises to let you hear just the sounds from the headphones.

Enabling subtitles

The subtitles feature overlays text transcriptions — which Amazon calls *subtitles* — of the voice track in a TV show, movie, or video. Follow these steps to enable and customize subtitles:

1. **Choose Settings ⇨ Accessibility ⇨ Subtitles.**

2. **Choose the Subtitles setting.**

   Fire TV turns subtitles on and displays extra settings that enable you to customize the appearance of the subtitles. The phrase *This is a preview* that appears on the right of the screen shows you what your custom subtitles will look like. The main settings are as follows:

- **Text:** Changes the text size, color, and font, among others.

- **Text Background:** Changes the color of the background on which the subtitles appear.

- **Window Background:** Changes the color of the background of the entire horizontal strip on which the subtitles and their background appear.

TIP

For a bit more control over the look of your subtitles as they appear in Amazon videos, send your nearest web browser to www.primevideo.com/cc and use the Subtitles tab to configure your subtitle presets. Back on the Fire TV Subtitles screen, make sure the Use Amazon Web Settings for Amazon Video Subtitles setting is set to On.

3. **For each setting you want to customize, choose the setting, use the screen that shows up to adjust the setting to your liking, and then press Back to return to the settings.**

TIP

If you end up with subtitles that look downright awful, you can start over again quickly by scrolling to the bottom of the Subtitles screen and then choosing the Reset to Defaults command.

You can also control the look of subtitles as you're watching a movie or TV show. During playback, press Menu to display the show's option, and then choose Subtitles (or, in some cases, Subtitles and Audio). Fire TV displays controls similar to the ones shown in Figure 8-10, which enable you to choose a language, type size, and subtitle style. Note that you only see the Size and Style options if you have subtitles turned on in Settings.

FIGURE 8-10:
You can use these controls to customize subtitles as you watch a show.

REMEMBER

Not all videos support subtitles. When you open the details screen for a TV show or movie, look for the CC label, which identifies shows that support closed captioning (subtitles).

Browsing the Web

Your big-screen TV is awfully nice for watching your favorite TV shows and the latest movies, so wouldn't it be even nicer if you could also use that screen real estate for your *other* favorite pastime? I speak, of course, of surfing the web and all that entails, from searching to shopping to social media. You bet that would be nice, and it's eminently doable thanks to Fire TV support for not one, but *two* web browsers:

» **Amazon Silk:** This is Amazon's Fire TV browser. It has a few video-related features, but it's easily used as a relatively full-featured browser.

» **Mozilla Firefox:** This is the Fire TV version of the popular Firefox web browser, and it's optimized for viewing and locating video content.

To install a web browser, choose Apps ⇨ Categories ⇨ Web Browser, choose the browser you want to use, and then choose Download.

Surfing with Silk

When you first open Silk, you see a Home screen similar to the one shown in Figure 8-11.

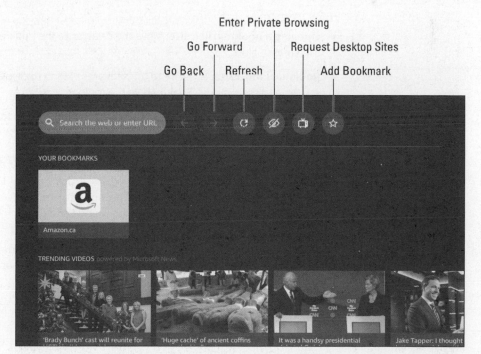

FIGURE 8-11:
The Home screen of the Silk web browser.

There are four main sections on the Home screen:

>> **Toolbar:** Icons for controlling Silk and page navigation:

- **Search the Web or Enter URL:** Enter search text or an address for the page you want to view.

- **Go Back:** Go back in the history of the current surfing session to the most recent page you've visited.

- **Go Forward:** Go forward in the history of the current surfing session to the most recent page you've visited.

- **Refresh:** Ask the web server to reload the current page.

- **Enter Private Browsing:** Run Silk in Private Browsing mode, which doesn't save page browsing data such as the history of sites you visit and your search requests.

- **Request Desktop Sites:** Request the regular (desktop) version of the page rather than the default mobile version.

- **Add Bookmark:** Create a bookmark for the current page, which then appears in the Your Bookmarks section (see the next bullet point) for fast access.

>> **Your Bookmarks:** Pages you've bookmarked for easy access. Note that Silk includes a bookmark for Amazon by default. Here are the techniques you can use to work with bookmarks:

- **Opening a bookmarked site:** Move the selector to the bookmark and then press Select.

- **Bookmarking a site:** Navigate to the site, press Menu to display the Home screen, and then choose Bookmark in the toolbar.

- **Deleting a bookmark:** On the Home screen, move the selector to the bookmark, press Menu, and then choose Delete.

>> **Trending Videos:** A selection of videos curated by Microsoft News.

>> **Manage:** Displays several tiles for managing Silk, including Clear Data (delete all your saved Silk browsing history, site data, and site files), Cursor Speed (change the speed of the cursor and scrolling), and Settings.

TIP

One setting you may want to change is the default search engine, which is Bing. To change the search engine, choose Settings ➪ Advanced ➪ Search Engine, and then select the search engine you prefer to use.

After you've used Silk to navigate to a web page, you see a large circle in the middle of the screen (pointed out in Figure 8-12), which Silk refers to as the *cursor*. You have two ways to manipulate the cursor:

>> Use the Fire TV remote's navigation buttons to move the cursor over the web page element you want to work with, such as a link or a text box. Press Select to choose that element (for example, open the linked page or enter text into the text box).

>> Use a connected Bluetooth or USB mouse to move the cursor and click links and other page elements.

To return to the Silk Home screen, press Menu.

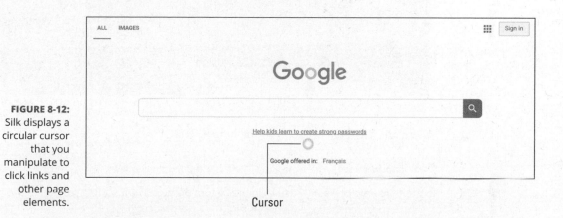

FIGURE 8-12: Silk displays a circular cursor that you manipulate to click links and other page elements.

Surfing with Firefox

When you first launch Firefox, the app asks if you want to leave Turbo mode turned on. Turbo mode automatically blocks trackers and ads, so it's best to leave Turbo mode turned on. If you find that Turbo mode causes a website to break in some way, you can always turn off Turbo mode (as I describe in the following list) while viewing that site.

When Firefox opens, you see a screen similar to the one shown in Figure 8-13, which Firefox calls the *homescreen*.

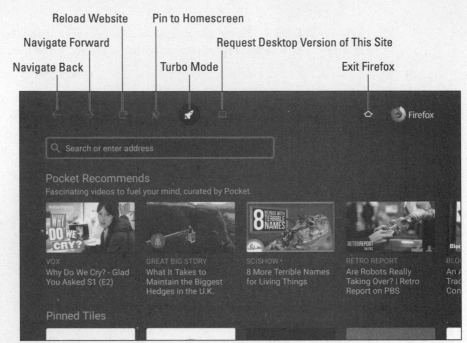

Navigate Back Navigate Forward Reload Website Pin to Homescreen Turbo Mode Request Desktop Version of This Site Exit Firefox

FIGURE 8-13:
The homescreen of the Firefox Fire TV web browser.

There are five main sections on the homescreen:

>> **Toolbar:** Icons for controlling Firefox and page navigation:

  • **Navigate Back:** Go back in the history of the current surfing session to the most recent page you've visited.

  • **Navigate Forward:** Go forward in the history of the current surfing session to the most recent page you've visited.

  • **Reload Website:** Ask the web server to reload the current page.

  • **Pin to Homescreen:** Add a tile for the current web page to the Pinned Tiles section of the homescreen.

  • **Turbo Mode:** Toggle Turbo mode off or on.

  • **Request Desktop Version of This Site:** Ask the web server to display the regular (desktop) version of the page rather than the default mobile version.

  • **Exit Firefox:** Quit Firefox and return to Fire TV.

>> **Search or enter address:** Displays the address of the current page and enables you to enter a search request or an address for the page you want to view.

- » **Pocket Recommends:** A selection of videos curated by Pocket, a Mozilla company (the same company that makes Firefox).

- » **Pinned Tiles:** Easy access to a selection of video-related web pages. Here are the techniques you can use to work with pinned tiles:

    - **Opening a pinned site:** Move the selector to the pinned tile and then press Select.

    - **Pinning a site:** Surf to the site, press Menu to display the homescreen, and then choose Pin to Homescreen in the toolbar.

    - **Unpinning a site:** In the homescreen, move the selector to the tile of the site, press and hold the Select button, and then choose Remove.

- » **Settings:** Enables you to configure a limited number of Firefox settings.

After you've surfed your merry way to a page, you see a large circle in the middle of the screen (similar to the Silk cursor pointed out in Figure 8-12), which is the pointer. You have two ways to manipulate the pointer:

- » Use the Fire TV remote's navigation buttons to move the pointer over the web page element you want to work with, such as a link or a text box. Press Select to choose that element (for example, open the linked page or enter text into the text box).

- » Use a connected Bluetooth or USB mouse to move the pointer and click links and other page elements.

To return to the Firefox homescreen, press Menu.

Listening to Music

If you have an awesome sound system connected to your Fire TV device — or, even better, if your Fire TV device *is* an awesome sound system, such as a Fire TV Edition Soundbar — then cranking out tunes at an acceptably high volume is a must.

Although Fire TV is mostly about video stuff, it's on speaking terms with a ton of audio content via apps such as Spotify and TuneIn Radio. To see and install a music or audio app, choose the Apps tab, choose Categories, and then open Music & Audio.

If you're an Amazon Prime member, however, you probably want to get your music through Amazon, which offers two plans:

» **Prime Music:** This is free to Amazon Prime members and offers about two million songs ad-free.

» **Amazon Music Unlimited:** This costs Amazon Prime members $7.99 per month (or $79 per year) or non-Prime member $9.99 per month, but you get access (according to Amazon) to "tens of millions" of songs (and it's ad-free).

Touring the Amazon Music app

Whichever plan you go with, you use the Amazon Music app on Fire TV to access your music and play it through your TV or its connected sound system. When you launch Amazon Music, you end up on a screen similar to the one shown in Figure 8-14.

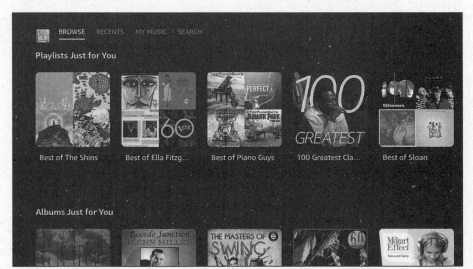

FIGURE 8-14:
The home screen of the Amazon Music app.

You navigate Amazon Music using the items displayed along the top of the screen:

» **Now Playing:** A thumbnail of whatever is currently playing. Choose the thumbnail to return to the current song.

- ➤ **Browse:** Displays a collection of playlists, albums, songs, and stations, many of which have been curated based on the music you've played in the past.

- ➤ **Recents:** Displays the albums, playlists, and stations that you've listened to most recently.

- ➤ **My Music:** Displays the artists, albums, playlists, genres, and songs that you've added to your My Music list on the Amazon Music website (`http://music.amazon.com`).

- ➤ **Search:** Enables you to search for an artist, album, song, playlist, or station.

Controlling music playback

Use your Alexa Voice Remote (or the remote feature in the Fire TV mobile app) to control playback using the following buttons:

- ➤ **Play/Pause:** Pause and restart the song.

- ➤ **Rewind:** Skip back to the beginning of the song. Press Rewind twice to skip to the previous song in the current album or playlist.

- ➤ **Fast Forward:** Skip to the next song in the current album or playlist.

- ➤ **Volume Up:** Raise the volume one notch; press and hold Volume Up to raise the volume quickly.

- ➤ **Volume Down:** Decrease the volume by one value; press and hold Volume Down to decrease the volume quickly.

- ➤ **Mute:** Toggle the volume off and on.

In the playback screen, shown in Figure 8-15, you can also access the following controls:

- ➤ **Repeat:** Choose this icon to start the current album or playlist over from the beginning when it ends. Choose this icon a second time to repeat only the currently playing song.

- ➤ **Shuffle:** Choose this icon to play the songs in the current album or playlist in random order.

- ➤ **More Options:** Choose this icon to display two extra controls: View Artist (to see more music from the current artist) and View Album (to see a list of the songs on the current album).

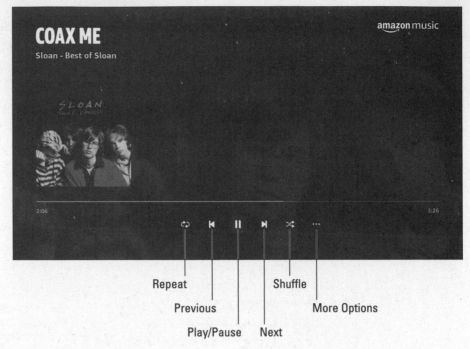

FIGURE 8-15:
The playback
screen of the
Amazon
Music app.

Repeat

Previous

Play/Pause

Shuffle

More Options

Next

For voice control of the music playback, here's a sampling of what you can say:

- "Pause."
- "Resume."
- "Previous."
- "Next."
- "Volume up."
- "Volume down."
- "Set volume to X" (where X is the volume setting you want).
- "Mute."
- "Unmute."
- "Turn shuffle on."
- "Turn shuffle off."
- "Turn repeat on."

>> "Turn repeat on song."

>> "Turn repeat off."

REMEMBER

In the preceding playback requests, if you're using a Fire TV Cube or other Alexa-enabled device, remember to start the request with the wake word *Alexa*.

Viewing Your Photos and Videos

You'll spend the bulk of your Fire TV career watching content produced by other people, but that doesn't mean you can't watch your own stuff, and by "stuff" I mean your own photos and videos. There are two roads you can take to viewing personal content on Fire TV:

>> **Media Player:** On Fire TV Edition televisions, you can use this app to view photos and videos that are stored on an external USB drive. As a bonus, Media Player is also happy to play any music files that you add to the drive. Note that, for reasons unknown, Fire TV also refers to this app as Gallery.

>> **Amazon Photos:** If you're an Amazon Prime member, you get free unlimited storage of your photos and videos on Amazon Photos, which you can then access through Fire TV using the Amazon Photos app.

Viewing photos and videos using Fire TV Edition Media Player

Earlier in this chapter (see "Adding a USB storage drive"), I mention that if you ask Amazon to format the drive as an external device, then you can use the drive to store files such as photos and videos (and music files, too). After you've added your files to the drive, reconnect it to your Fire TV device, wait until you see the notification shown in Figure 8-16, and then press Menu to launch the Media Player app.

FIGURE 8-16:
The home screen of the Amazon Music app.

View Pictures
Press menu to view pictures on your USB drive
Launch

TIP

The notification only appears onscreen for a few seconds. If you miss it, there are a couple of routes you can take to launch Media Player for the first time:

>> Choose Settings ⇨ Inputs ⇨ Media Player.

>> Choose Settings ⇨ Applications ⇨ Manage Installed Applications ⇨ Gallery ⇨ Launch Application.

The first time you launch Media Player (or Gallery, as Fire TV often insists on calling it), the app asks for permission to access the files on your device, as shown in Figure 8-17. Choose Allow.

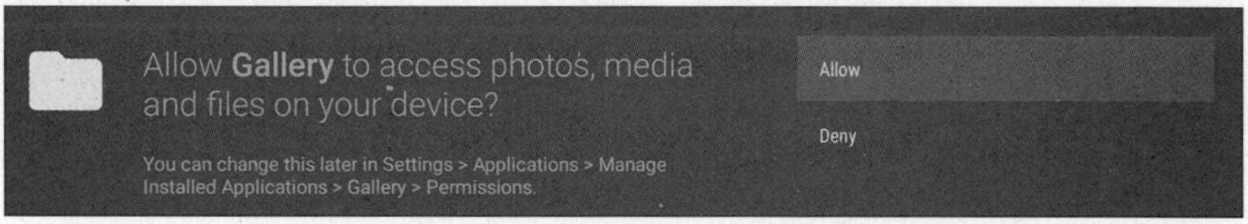

FIGURE 8-17:
When Media Player asks for permission to access your files, choose Allow.

Allow **Gallery** to access photos, media and files on your device?

You can change this later in Settings > Applications > Manage Installed Applications > Gallery > Permissions.

Allow

Deny

When Media Player loads, it examines your external USB drive for photos, videos, and music files. Note that it looks both in the main folder of the drive and in any subfolders that the drive contains. You navigate Media Player using the following four tabs at the top of the screen:

>> **Your Files:** Displays all the media files and subfolders found on the drive.

>> **Videos:** Displays all the video files and subfolders that contain video files found on the drive.

>> **Images:** Displays all the image files and subfolders that contain image files found on the drive.

>> **Audio:** Displays all the audio files and subfolders that contain audio files found on the drive.

When you navigate to a tab, Media Player shows several subtabs — View All, 0, Photos, and Seagate USB Drive — pointed out in Figure 8-18. Choose View All to see everything; choose 0 to see what's stored internally on Fire TV; and choose the other subtabs to see what's in the drive's main folder and subfolders.

Choose an item to display it; then use the Right and Left navigation ring buttons to cycle through the media.

Subtabs

FIGURE 8-18: Selecting a tab displays several subtabs.

Viewing photos and videos with the Amazon Photos app

If you're an Amazon Prime member who has taken advantage of the unlimited storage you get with Amazon Photos, you may wonder if you can then view those photos on Fire TV. Sure you can! Choose Apps ➪ Categories ➪ Photo & Video, and then install the Amazon Photos app.

When you first launch Amazon Photos, the app takes you through a brief introduction. With the intro out of the way, the main app screen displays a long line of tabs across the top: Your Photos, Family Vault, Videos, Albums, Folders, People, and Places. Select a tab to view the photos (or videos) in that category; then choose a photo to view it full-screen. Use the Right and Left navigation ring buttons to move through the photos. When you're done, press Back to return to the previous tab.

Running a Fire TV slideshow

Notice in Figure 8-18 that the upper-right corner of the screen includes a Start Slideshow command. The same command also appears in Amazon Photos when you open a photo location. Choose Start Slideshow (or press Play/Pause) to display the current folder's photos in a slideshow on Fire TV.

During the slideshow, press Menu to set the following options:

>> **Album:** Choose a different photo location on the external drive.

>> **Slide Style:** Choose how you want Fire TV to display each image (such as Pan & Zoom or Dissolve).

>> **Slide Speed:** Choose the rate at which Fire TV displays the photos: Slow, Medium, or Fast.

>> **Shuffle:** Toggle shuffle on (to display the photos randomly) or off (to display them in the order in which they appear in the folder).

Fire TV also displays a slideshow as a screensaver when your Fire TV has been idle for a set amount of time. To configure this screensaver, choose Settings ⇨ Display & Sounds ⇨ Screen Saver. Besides the options I list earlier, you can also configure the following:

>> **Current Screensaver:** You can pull screensaver photos from Amazon, your most recent photos added to Amazon Photos, or from your Photo Booth folder.

To set a specific Amazon Photos folder as the screensaver source, open the folder, press Menu, and then choose Set as Screensaver. As useful as it may seem, I'm afraid that the Fire TV screensaver doesn't offer the option of using a folder from your external storage drive.

>> **Start Time:** Specify how much time Fire TV must be idle before the screensaver kicks in.

>> **Alexa Hints:** Toggles on or off the screensaver text that gives you hints on how to use Alexa.

Chapter **9**

Controlling Fire TV (And More) with Alexa

Whatever Fire TV device you own, if you're like most people, you probably think that what Fire TV brings to your home is an easily configured and navigable streaming media device. That's certainly true, but Fire TV comes with a "bonus" feature that lots of people miss: It turns your TV into a smart speaker, which means it can listen for, understand, and carry out your voice instructions or questions. The magic behind this wondrous capability is Alexa, Amazon's powerful and popular voice service. All Fire TV devices are Alexa-friendly, which means you get access to the world of voice control of Fire TV devices without needing to buy anything extra. Even better, you can also use your Alexa-enabled Fire TV device to dive into the fascinating realm of smart-home automation, which enables you to use voice instructions to control lights, turn devices on and off, and much more. In this chapter, you explore the connection between Fire TV and Alexa.

What Is Alexa?

The basic premise of Alexa couldn't be simpler: You ask a question, Alexa provides the answer; you give a request, Alexa carries it out. However, this surface simplicity lies on top of a mind-bogglingly complex system that involves hardware, software, networking, artificial intelligence, and the cloud. Fortunately, that underlying complexity isn't something you have to deal with.

Why is Alexa so popular? There are lots of reasons, but the one that really matters is that Alexa is (or tries hard to be) a "voice service." Older voice-command tools were geared toward using a computer: running programs, pulling down menus, accessing settings, and so on. Alexa doesn't do any of that. Instead, it's focused on doing things for you in your real life, including (but by no means limited to) the following:

>> Locating and playing movies and TV shows

>> Playing music, podcasts, or audiobooks

>> Setting timers and alarms

>> Telling you the latest news, weather, or traffic

>> Creating to-do lists and shopping lists

>> Buying stuff from Amazon

>> Controlling smart-home devices such as lights and thermostats

Alexa's components

Throughout this chapter I talk about "Alexa" as though it's a single object, but Alexa is really a large collection of objects that, together, create the full, seamless Alexa experience. For the purposes of this chapter, Alexa consists of the following four components:

>> **Name recognition:** Although when you interact with Alexa it seems as though the device understands what you say, the only speech your Alexa device actually recognizes is the word *Alexa,* which Amazon calls the *wake word.* That is, it's the word that lets Alexa know it should wake up and start listening for an incoming request or question.

>> **Speech recording:** Your Alexa device (such as the Alexa Voice Remote or the Fire TV Cube) has a built-in microphone that captures the questions, requests, and other utterances that you direct to the device. There's a simple computer

inside that records what you say and then sends the recording over the Internet to the Alexa Voice Service (discussed next).

» **Alexa Voice Service (AVS):** Here's where the real Alexa magic happens. This part of Alexa resides in Amazon's cloud. AVS takes the recording that contains your voice request and uses some fancy-schmancy speech recognition to tease out the actual words you spoke. AVS then uses natural-language processing to analyze the meaning of your request, from which it produces a result.

» **Speech synthesis:** This component takes the results provided by AVS and renders them as speech, which is stored in an audio file. That file is returned and played through the Alexa device's built-in or connected speakers.

How Alexa works

Given the various Alexa components that I outline in the preceding section, here's the general procedure that happens when you interact with Alexa to get something done:

1. **You say "Alexa" (if you have a Fire TV Cube or Echo smart speaker) or you press the Voice button on the Alexa Voice Remote.**

Your Alexa device wakes up and begins hanging on your every word.

2. **You state your business: a question, a request, or whatever.**

The Alexa devices records what you say. When you're done, the device uses your Internet-connected Wi-Fi network to send the recording to AVS in Amazon's cloud.

WARNING

It may seem like Alexa "lives" inside whatever device you have, but Alexa is very much an Internet-based service. That means if you don't have Internet access, you don't have Alexa access either.

3. **AVS uses its speech-recognition component to turn the recorded words into actual data that can be analyzed.**

4. **AVS uses its natural-language processing component to analyze the words in your request and then figure out exactly what you asked Alexa to do.**

AVS doesn't analyze every single word you say. Instead, it's mostly looking for the telltale *keywords* that indicate what you've asked Alexa to provide. For example, if you said, "What's the weather forecast for tomorrow?," all AVS needs are the words *weather, forecast,* and *tomorrow* to deliver the correct info.

5. **If AVS can't fulfill your request directly, it passes the request along to a third-party service (such as AccuWeather or Wikipedia), and then gathers the response.**

6. **AVS returns the response via the Internet to your Alexa device.**

   What AVS returns to your Alexa device depends on the result. If the result is just information for you, AVS converts the info to speech and stores the speech in an audio file that your Alexa device can play. If the result is a request (for example, to play a particular song), AVS passes that request back to the Alexa device.

7. **The Alexa device either uses its built-in or connected speaker to broadcast the result of your request or carries out your request.**

   You also see the result on your Fire TV screen.

Installing the Alexa App

The Alexa app is a program that you download to your smartphone or tablet. With the Alexa app, you can connect your Alexa devices to your Wi-Fi network, provide Alexa with your location, connect to smart-home devices, provide the details of your Amazon account, and much more. You *use* Alexa by conversing with an Alexa-friendly device, but you *configure* Alexa using the app.

Okay, so what do you need to get the Alexa app? Either of the following:

>> **A smartphone or tablet that meets one of these qualifications:**

  - An iPhone or iPad running iOS 11 or later

  - An Android phone or tablet running Android 5.1 or later

  - An Amazon Fire tablet or phone running Fire OS 5.5.3 or later

  If you have one of these devices, go to your device's app store, search for the Alexa app, and install it.

  If you have one of Amazon's Fire devices and that tablet is capable of running Alexa, the Alexa app will be installed automatically on the device.

REMEMBER

>> **A Windows or Mac web browser:** In this case, point your browser to `https://alexa.amazon.com`. This takes you to the Amazon Alexa portal page.

When you first open the Alexa app or surf to the Amazon Alexa portal page, you're prompted to sign in with your Amazon account credentials. Go ahead and sign in, and then follow the onscreen instructions to get the app configured. When you're done, you see the Home screen, which will be similar to the one shown in Figure 9-1.

WARNING

For best results, make sure you sign in to the Alexa app using the same Amazon account as you use with the Fire TV mobile app and your Fire TV device.

Menu

FIGURE 9-1:
The Alexa app's
Home screen.

Home   Alexa   Play Devices

Communicate

Taking a tour of the Alexa app

In Figure 9-1, I point out a few landmarks of the Alexa app screen. Here's a summary:

>> **Menu:** Tap this icon to access the main app menu, shown in Figure 9-2. You use the requests on this menu to configure various Alexa features (such as music and reminders), add and manage Alexa skills, and change the Alexa app's settings.

Add Device

Lists

Reminders & Alarms

Contacts

Routines

Things to Try

Skills & Games

Activity

Help & Feedback

Settings

FIGURE 9-2:
Tap the Menu
icon to see this
menu of app
requests.

>> **Home:** Tap this icon to display the Home screen, which offers the Things to Try section (suggestions for getting started with Alexa) and (eventually) a series of sections — known as *cards* — that present in reverse chronological order (that is, the newest at the top) your most recent interactions with Alexa: your questions and Alexa's answers, responses to your requests, recently played music and other media, and more.

>> **Communicate:** Tap this icon to open the Communication screen. Here, you can set up and work with Alexa's communications features, including starting a new call, sending a text message, and dropping in on someone.

>> **Alexa:** Tap this icon to send questions and requests to Alexa via your smartphone or tablet microphone. See "Giving Alexa access to your device microphone," next.

>> **Play:** Tap this icon to see what media — such as a song, podcast, or audiobook — you've played recently on your Alexa device. You can also browse your Audible and Kindle libraries.

>> **Devices:** Tap this icon to work with both your Alexa devices and Alexa's smart-home features. You can add, configure, and operate devices, manage smart-home skills, create smart-home groups, and more.

Giving Alexa access to your device microphone

One of the nice features of the Alexa app is that you can use it to send voice requests to Alexa via your smartphone or tablet microphone. That's the purpose of the Alexa button in the middle of the Alexa app toolbar at the bottom of the screen (refer to Figure 9-1). However, before you can send voice requests to Alexa, you need to give the Alexa app permission to use your device's microphone. Here's how it works:

1. In the Alexa app, tap the Alexa icon (refer to Figure 9-1).

You see some text telling you that you need to give the app permission to use the microphone.

2. Tap Allow.

Your device asks you to confirm.

3. Tap OK (iOS) or Allow (Android).

Your device gives the Alexa app permission to access the microphone and then displays some examples of things you can say to Alexa.

If you also see a prompt asking you to give the Alexa app permission to use your location, go ahead and tap Allow.

4. Tap Done.

You now hear a tone and see the regular Alexa screen, which is waiting for you to say something to Alexa. You'll want to skip that for now, so either tap the X to close the screen or wait a few seconds for the screen to close by itself.

Connecting Your Alexa Device to Fire TV

To set up your Alexa device to control your Fire TV device, you have to introduce them to each other. Here's how it's done:

1. In the Alexa app, tap Menu ⇨ Settings.

2. Tap TV & Video.

3. Tap Fire TV.

4. Tap Link Your Alexa Device.

The Alexa app displays a list of your Fire TV devices.

5. **Select the Fire TV device you want to control (see Figure 9-3), and then tap Continue.**

   The Alexa app displays a list of your Alexa devices.

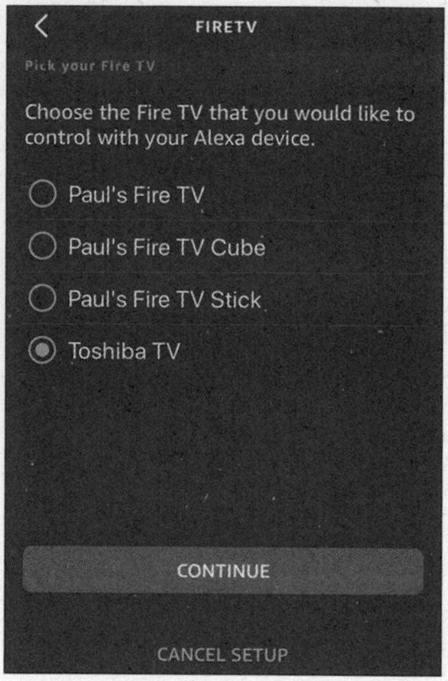

FIGURE 9-3:
Tap the Fire TV
device you want
to control with
Alexa.

6. **Select the Alexa device you want to use to control the Fire TV device you chose in step 5.**

   If you have multiple Alexa devices, tap each device that you want to allow to control your Fire TV device.

7. **Tap Link Devices.**

   The Alexa app connects your Alexa device and your Fire TV device.

Getting to Know Alexa

Alexa is designed to be as simple as possible and to have a very slight learning curve, especially at the start. That's good news for anyone just beginning with Alexa because it means you have to learn only a few basics. The next few sections walk you through what you need to know.

Getting Alexa's attention

It may seem as though, when it's not in use, Alexa is just sitting there listening to you talk to yourself, but that's not really the case. Instead, it's more accurate to imagine Alexa spending its off time in a light, dreamless slumber where it has no idea what's happening around it. Your job is to interrupt that slumber by gently tapping Alexa to rouse it and get its attention.

There are several ways to get Alexa's attention, but here are the most common:

>> On the Alexa Voice Remote, press and hold the Voice button.

>> On the Fire TV mobile app, pull down and hold the Voice icon.

>> For Fire TV Cube or an Alexa-enabled device, such as an Echo smart speaker, say, "Alexa."

>> For Fire TV Cube or an Alexa-enabled device, press the Action button.

>> In the Alexa app, tap the Alexa button.

Keeping Alexa's attention

Most of the time, you interact with Alexa by issuing intermittent voice requests: You ask for the time now, the temperature a minute later, and whether Dustin Hoffman was in *Star Wars* a few minutes after that. However, every now and then you may want to issue a series of requests, one after the other. That's perfectly fine, but it gets a bit old having to say, "Alexa" at the start of each voice request.

Forget that. Instead, you can put Alexa into Follow-Up mode, which enables you to say the wake word once and then issue multiple requests without having to say the wake word again.

WARNING

Follow-Up mode is slick, but I should warn you that it doesn't always work. That is, Alexa won't switch into Follow-Up mode in these situations:

>> When you've forced Alexa to stop responding by issuing a request to end the conversation, which I describe in the next section

>> When your Alexa device is playing media (such as a song or audiobook)

>> When Alexa can't be certain that you're speaking to it and not to someone else nearby

Here are the steps to run through to enable Follow-Up mode:

1. **In the Alexa app, choose Devices ⇨ Echo & Alexa.**

   The Echo & Alexa screen appears with a list of your Alexa devices.

2. **Tap the device you want to work with.**

   The Device Settings screen appears.

3. **Tap Follow-Up Mode.**

4. **Tap the Follow-Up Mode setting to On, as shown in Figure 9-4.**

   The Alexa app updates the device with the new setting.

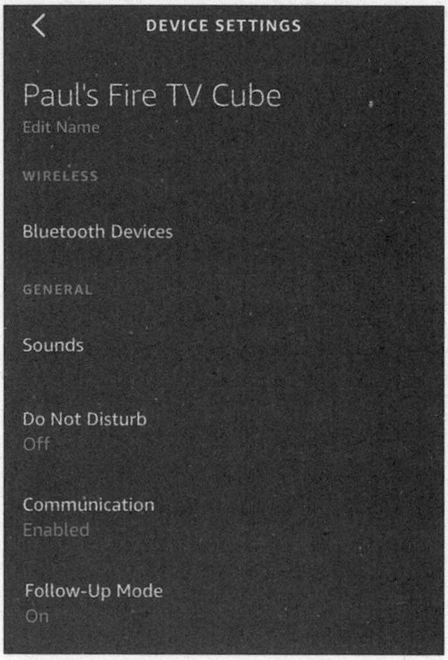

FIGURE 9-4:
Tap the Follow-Up
Mode switch to
On to enable
Follow-Up mode.

Ending the conversation

Alexa, bless its digital, cloud-based heart, can be very helpful, but sometimes it's *too* helpful. That is, you can ask Alexa a question and still be listening to the answer a minute later! If you find that Alexa is going on and on about something, you can end the monologue (Amazon calls this "ending the conversation" — ha!) by interrupting Alexa with any of the following requests:

>> "Stop."

>> "Enough."

172   PART 3 Going Beyond the Basics

>> "Shush."

>> "Cancel."

>> "Thank you."

>> "Sleep."

Enabling Brief mode

Alexa's infamous long-windedness has long been a complaint of users, so Amazon finally did something about it: It created a feature called Brief mode. In Brief mode, Alexa gives shorter answers than usual, and in situations where it usually gives a quick "Okay" or similarly useless response, it now just plays a sound. If that sounds like bliss to you, follow these steps to turn on Brief mode:

1. **In the Alexa App, choose Home ⇨ Menu ⇨ Settings.**

2. **Tap Alexa Account.**

3. **Tap Alexa Voice Responses.**

4. **Tap the Brief Mode switch to On, as shown in Figure 9-5.**

   The app configures Alexa to use Brief mode.

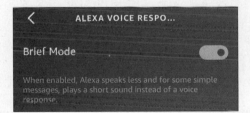

FIGURE 9-5:
Tap the Brief
Mode switch to
On to enable
Brief mode.

Enabling Whisper mode

If you have people nearby, the last thing they probably want to hear is both your Alexa utterances and Alexa's responses at full volume. Does that mean you have to forgo Alexa until you're alone? Not necessarily, you can enable Alexa's Whisper mode, which means that when you whisper a request to Alexa, Alexa whispers its response back. (Yes, this is very weird at first, even a bit creepy. Perhaps you'll get used to it!)

The easiest way to enable Whisper mode is with a voice request:

"Turn on Whisper mode."

Inanely, Alexa responds at full volume to tell you that Whisper mode has been activated! Alternatively, you can enable Whisper mode more quietly by using the Alexa app:

1. **In the Alexa app, choose Home ⇨ Menu ⇨ Settings.**

2. **Tap Alexa Account.**

3. **Tap Alexa Voice Responses.**

4. **Tap the Whisper Mode switch to On, as shown in Figure 9-6.**

   The app configures Alexa to use Whisper mode.

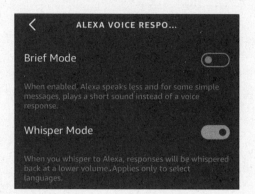

FIGURE 9-6: Tap Whisper Mode to On to enable Whisper mode.

Watching Movies and TV Shows with Alexa

Alexa can do an amazing range of things, but as a Fire TV user I'm sure your main concern is using Alexa to locate, play, and control the playback of movies and TV shows on your Fire TV device. In the next few sections, I go through a few ways to use voice requests to interact with movies and TV shows.

REMEMBER

With all the voice utterances I list in this chapter, remember that if you're interacting with Alexa through a Fire TV Cube, an Echo smart speaker, or a similar Alexa-enabled device, then you need to precede each utterance with the wake word *Alexa*.

Navigating Fire TV tabs

To get around the Fire TV interface, you can use these voice requests:

» "Go Home."

» "Go to Live."

» "Go to Your Videos" (or "My Videos" or "Videos").

» "Go to DVR" (only if you have Fire TV Recast installed).

» "Go to Movies."

» "Go to TV Shows."

» "Go to Apps."

» "Go to Settings."

Locating movies and TV shows

You can use any of the following voice requests to search for shows on Fire TV:

» "Show me the movie [*movie title*]."

» "Show me the TV show [*series title*]."

» "Show me [*actor*] movies."

» "Show me [*genre*] movies."

» "Show me [*genre*] TV shows."

In each case, you can replace "Show me" with either "Search for" or "Find."

Watching a movie trailer

If you want help deciding whether an upcoming or recently released movie is something you'd like to check out, or if you just get a kick out of previews, you can ask Alexa to play you the trailer for a movie. Use either of the following voice requests:

» "Play the trailer for [*movie title*]."

» "Show the trailer for [*movie title*]."

Playing movies and TV shows

After you have a movie or TV show playing, you can also use the following voice requests to control the playback:

» "Pause."

» "Resume" (or "Play").

» "Rewind" (rewinds 10 seconds).

» "Rewind *X* seconds."

» "Fast-forward" (fast-forwards 10 seconds).

» "Fast-forward *X* seconds."

» "Next episode."

» "Previous episode."

Watching live TV

If your Fire TV includes live TV channels via an HDTV antenna connection or other source, you can use the following voice requests to display the channel guide:

» "Channel guide."

» "Open channel guide."

» "Show me the channel guide."

You can also use the following voice requests to tune to a station from anywhere in the Fire TV interface:

» "Change to [*channel* or *network*]."

» "Go to [*channel* or *network*]."

» "Tune to [*channel* or *network*]."

» "Watch [*channel* or *network*]."

In these requests, replace *channel* with the station's channel number (for example, "Tune to 5.1") or replace *network* with the station's network name (for example, "Watch NBC").

How do you know the station's channel number? You have three ways to find out:

>> Choose Settings ⇨ Live TV ⇨ Channel Management, and then choose your live TV source (such as Antenna Channels). In the list of channels that appears, each station displays its channel number.

>> On your Fire TV Edition version of the Alexa Voice Remote, press the Guide button to display the channel guide. As you scroll vertically through the guide, you see the channel number for each station below the station logo.

>> Use the TV Fool website (www.tvfool.com) to search for broadcast stations in your area, as I describe in Chapter 10. In the TV Fool results, the channel numbers you want are listed in the Channel section's (Virt) column.

Controlling the volume

To get the playback volume just right, here are some requests you can use:

>> "Volume up."

>> "Increase the volume."

>> "Raise the volume."

>> "Louder."

>> "Volume down."

>> "Decrease the volume."

>> "Lower the volume."

>> "Softer."

>> "Set volume to X" (where X is the volume setting you want).

>> "Volume X" (where X is the volume setting you want).

>> "Mute."

>> "Unmute."

Movie and TV show info requests

Here are a few useful requests related to getting information about TV shows and movies:

>> "What is the IMDb rating for [TV show or movie title]?"

>> "Tell me about the movie [movie title]."

>> "When was movie [*movie title*] released?"

>> "Who stars in [*TV show or movie title*]?"

>> "What is [*actor*]'s latest movie?"

>> "Which movie won the Best Picture Oscar in [*year*]?"

>> "How many Oscars has [*actor*] won?"

Some movie Easter eggs

In the software world, an *Easter egg* is a whimsical program feature that's hidden by default and must be discovered. Alexa, you'll be delighted to know, contains hundreds of Easter eggs. You can get a random one using either of these requests:

>> "Give me an Easter egg."

>> "Give me a hard-boiled Easter egg."

For a more targeted Easter egg hunt, you can try specific topics such as TV shows, which I discuss in the next section. Movies are another great source for Alexa Easter eggs. There are dozens, perhaps even hundreds, so consider the following ten to be a mere appetizer:

>> "Open the pod bay doors." *(2001: A Space Odyssey)*

>> "Who's on first?" (Abbot and Costello)

>> "Release the Kraken!" *(Clash of the Titans)*

>> "Klattu barada nikto." *(The Day the Earth Stood Still)*

>> "E.T. phone home." *(E.T. the Extra-Terrestrial)*

>> "Define supercalifragilisticexpialidocious" *(Mary Poppins)*

>> "Are we in the Matrix?" *(The Matrix)*

>> "What is my mission?" *(Mission: Impossible)*

>> "What is the airspeed velocity of an unladen swallow?" *(Monty Python and the Holy Grail)*

>> "What is the Jedi code?" *(Star Wars)*

Some TV Easter eggs

Alexa doesn't have a ton of TV Easter eggs, but here are a few to wet your whistle:

>> "What is your cunning plan?" *(Blackadder)*

>> "Who shot J.R.?" *(Dallas)*

>> "Don't mention the war." *(Fawlty Towers)*

>> "Winter is coming." *(Game of Thrones)*

>> "Who loves orange soda?" *(Kenan & Kel)*

>> "Who loves ya baby?" *(Kojak)*

>> "This is a dead parrot." *(Monty Python)*

>> "More cowbell." *(Saturday Night Live)*

>> "What is the Prime Directive?" *(Star Trek)*

>> "Where's the beef?" (Wendy's commercial)

More Useful Alexa Requests

Besides controlling movies and TV shows on your Fire TV, you can also use Alexa for non-media stuff, such as getting the current time and weather. The next few sections take you through a few dozen of Alexa's most common and most useful voice requests.

Everyday-info requests

Alexa excels at everyday requests for information related to the weather, news, and traffic. Here are the basic requests:

>> "What time is it?"

>> "What time is it in [*city*]?"

>> "What is today's date?"

>> "When is [*holiday*] this year?"

>> "What's the weather like?"

>> "What's the temperature?"

>> "Will it rain today?"

>> "What will the weather be like [*day*]?" (For example, "what will the weather be like tomorrow?")

>> "Play my flash briefing."

>> "What's in the news?"

>> "How's the traffic?"

>> "What was the score of yesterday's [*team*] game?"

>> "When is the next [*team*] game?"

>> "What movies are playing?"

>> "Find me a nearby [*cuisine*] restaurant." (For example, "Find me a nearby Mexican restaurant.")

Information requests

Give these requests a whirl to get general information from Alexa:

>> "What's the definition of [*word*]?"

>> "How do you spell [*word*]?"

>> "What are some synonyms for [*word*]?"

>> "Convert [*number*] [*units1*] to [*units2*]." (For example, "Convert 100 miles to kilometers.")

>> "What is [*number*] [*operator*] [*number*]?" (For example, "What is 123 times 456?")

>> Alexa, what is [*number1*] percent of [*number2*]?" (For example, "What is 15 percent of 78.53?")

>> "What is the population of [*city or region*]?"

Audio requests

Here are the basic voice requests to try out for controlling music, podcasts, and other audio:

>> "Play [*song or album*]."

>> "Play [*music genre*]."

- "Play [*playlist*]."
- "Play the latest [*artist*]."
- "Play Sunday morning music."
- "Who sings this song?"
- "Shuffle mode [*on or off*]."
- "Turn this off in [*number*] minutes."
- "Play [*audiobook title*]."
- "Read [*Kindle book title*]."

Alarm and timer requests

Here are the basic requests to use to set and work with alarms and timers:

- "Set alarm for [*time*]."
- "Wake me up every day at [*time*]."
- "Wake me up at [*time*] to [*music or radio station*]."
- "Snooze for [*number*] minutes."
- "Set a timer for [*number*] minutes."
- "Set a [*name*] timer for [*number*] minutes." (For example, "Set a bread timer for 40 minutes.")
- "How long is left on the timer?"
- "How long is left on the [*name*] timer?"
- "Stop the timer."

Calendar, reminder, and list requests

Here are a few basic Alexa requests for controlling your calendar, creating reminders, and managing lists:

- "What's on my calendar?"
- "What's on my calendar on [*date*]?"
- "Add an event to my calendar."

» "Create a new appointment."

» "Add budget meeting to my calendar on [*day*] at [*time*]."

» "What am I doing tomorrow?"

» "How many days until [*date*]?"

» "How many days until [*holiday*]?"

» "What time does the sun rise on [*date*]?"

» "What time does the sun set on [*date*]?"

» "Remind me to [*task*] in [*number*] minutes."

» "Add [*task*] to my to-do list."

» "What's on my to-do list?"

Communication requests

Here are the basic requests for placing calls and sending text messages:

» "Call [*name*]."

» "Call [*phone number*]."

» "Answer the call."

» "Hang up."

» "Drop in on the [*room or location*]."

» "Play messages."

» "Message [*name*]."

» "Announce that [*message*]."

Shopping requests

Here are a few requests to place and track Amazon orders:

» "Add [*item*] to my shopping list."

» "Buy more [*item*]."

» "Order [*item*]."

>> "What's on my shopping list?"

>> "Where's my stuff?"

>> "What are your deals?"

Using Alexa to Control Smart-Home Devices

Unless you're under 5 years old, you probably grew up in a home that was, well, dumb. You turned on lamps with a switch (or perhaps a clap), the thermostat told you only the current temperature, and the only task you could automate was setting your alarm clock. All that seemed perfectly normal at the time, but that dumb home is starting to look quaint when placed next to the modern idea of a smart home.

What, exactly, do people mean when they add the *smart* adjective to the word *home*? The simple — and not all that helpful — definition of a smart home is a home that contains one or more devices that enhance your home life in some way. That word *enhance* is vague, I know, but it's really the key to everything. How does a smart-home device enhance your home life? It comes down to three things:

>> **Convenience:** You operate most dumb-home devices manually, meaning you have to walk up to the device and then flip a switch or adjust a dial. If someone's at the door, the only way to see who's there is to open the door or peek through the peephole. By contrast, you operate smart-home devices from a distance using an app or a voice service such as Alexa. If someone's at the door, your smart security camera lets you see who's there using an app or an Alexa device with a screen.

>> **Information:** Dumb-home devices tell you either nothing about themselves or just the bare minimum. A dumb dimmer tells you nothing about the current light level, while a dumb thermostat shows only the current temperature. By contrast, smart-home devices are bristling with information — such as current settings, status indicators, and power usage — that gets relayed to an app or device for easy reference.

>> **Automation:** Dumb-home devices just sit there until you do something. A dumb lamp goes on when you flip the switch and will stay on until you flip the switch back. A dumb thermostat will keep the house at the temperature you set, no matter what the time of day. By contrast, smart-home devices can be

programmed. You can program a smart lamp to turn on and off automatically at specified times. You can program a smart thermostat to use your preferred temperature during the day, and to use an energy-saving lower or higher temperature overnight.

Yes, some smart-home stuff is a solution in search of a problem. A smart water bottle that tells you when it's time to take a drink and a smart hairbrush that lets you know when you're not brushing correctly are among the dumber smart devices. On the other hand, even something as basic (in the smart-home world, anyway) as programming when your lights go on and off can both save you money by reducing energy costs and extending bulb life, and make your home more secure by making it look occupied even when you're not there.

Installing a Wi-Fi smart-home device

If your smart-home device is Wi-Fi-friendly, go to your mobile device app store and install the manufacturer's app. Then follow these steps to get your Wi-Fi smart-home device set up in the app:

1. **Plug in and, if required, turn on the smart-home device.**

2. **Open the smart-home device manufacturer's app.**

3. **Initiate the procedure for setting up a new device.**

   Look for a request named Add or Add [*manufacturer*] Device (where *manufacturer* is the name of the company), or just a big plus sign (+).

   The setup routine will tell the device to broadcast its Wi-Fi network.

4. **Open your mobile device's Wi-Fi settings and look for the device's Wi-Fi network.**

   Figure 9-7 shows a collection of Wi-Fi networks that includes WeMo.Insight.03C, which is a network broadcast by a WeMo Insight smart switch.

5. **Tap the device network to connect to it.**

6. **When the connection is complete, return to the device app.**

   The app automatically detects the new network and uses the connection to set up the device. This usually involves giving the device a name. You'll often have to set up an account with the manufacturer, as well.

7. **The app will usually ask for your Wi-Fi credentials, which enable the device to connect to and operate over your network.**

   Having the device on your network is also how Alexa discovers and operates the device, so this step is important.

FIGURE 9-7:
Open your Wi-Fi settings and look for the smart-home device's Wi-Fi network.

8. **If you see a notice asking whether you want to upgrade the smart-home device firmware, by all means tap Yes or Allow or Update or whatever button answers in the affirmative.**

   The firmware is internal software that runs the device. Keeping all your smart-home devices updated with the latest firmware is highly recommended because new versions of the software are often needed to patch security holes and improve performance.

Making automatic network connections with Wi-Fi Simple Setup

The steps I outline in the preceding section mostly deal with getting a Wi-Fi-enabled smart-home device on your home network. The step where you need to connect your smartphone or tablet to the device's temporary network always bothers me because it seems like an imposition. The Amazon engineers must have felt the same way, because they came up with a way to avoid that annoying extra step. It's called Wi-Fi Simple Setup, and it requires two things:

» An Echo device compatible with Wi-Fi Simple Setup — that is, a second-generation or later Echo, Echo Plus, Echo Dot, or Echo Show — that's already connected to your Wi-Fi network

» The password to your Wi-Fi network saved to Amazon

If you've checked off both items, then setting up a new device that's compatible with Wi-Fi Simple Setup — such as the Amazon Smart Plug or the AmazonBasics Microwave — is either easy or ridiculously easy.

The ridiculously easy setup comes your way if you purchased your Wi-Fi Simple Setup device from Amazon. In that case, Amazon automatically associates the device with your Amazon account, which means that when you plug in the device, it will connect to your network automatically using your saved Wi-Fi password. Now *that's* ridiculously easy!

If you purchased the Wi-Fi Simple Setup device from a retailer other than Amazon, then the device won't be associated with your Amazon account, so it can't connect to your network automatically. That's okay, though, because you can still use the Alexa app to add the device: Choose Devices ⇨ Add (+) ⇨ Add Device.

Discovering smart-home devices using an Alexa skill

If you're not using an Alexa device that includes a smart-home hub (such as the Echo Plus and second-generation Echo Show), then you usually need to upgrade Alexa to work with your smart-home device. You upgrade Alexa by enabling the device manufacturer's Alexa skill. This not only lets Alexa discover the device, but also upgrades Alexa with the voice requests that let you operate the device.

REMEMBER

Alexa can locate and connect to some smart-home devices without requiring a skill. For example, Alexa can work with a Philips Hue Bridge to control lights without needing a skill. Say, "Discover devices," and then press the button on top of the Hue Bridge to put it into pairing mode.

Follow these steps to enable the manufacturer's Alexa skill and discover the manufacturer's smart-home device:

1. **Install the manufacturer's app and use it to get your smart-home device on your Wi-Fi network.**

   See "Installing a Wi-Fi smart-home device," earlier in this chapter, for the details.

2. **In the Alexa app, tap Devices.**

3. **Tap the Add button (+) that appears in the upper-right corner.**

4. **Tap Add Device.**

   The Alexa app displays icons for some popular brands and some device categories.

5. Tap the category that fits your device, and then tap the manufacturer.

The Alexa app prompts you to perform the duties I outline in Step 1. You've done all that, so proceed.

6. Tap Continue.

The Alexa app opens the information page for the manufacturer's Alexa skill.

7. Tap Enable.

At this point, what happens next depends on the skill, but you'll usually have to perform one or both of the following:

- Use the smart-home device app to give Alexa permission to access the device.

- Link Alexa to the user account associated with the smart-home device.

8. When you're done, tap Close (X) to return to the skill page.

9. Tap Discover Devices (see Figure 9-8).

The Alexa app uses the manufacturer's Alexa skill to search for available devices. If one or more devices are found, you see a screen similar to the one shown in Figure 9-9.

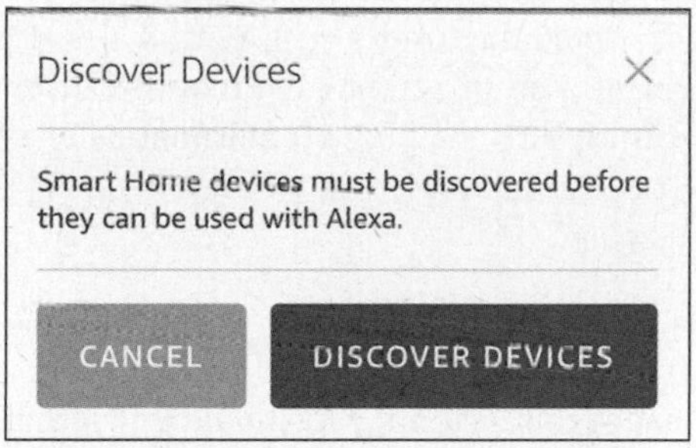

Discover Devices ✕

Smart Home devices must be discovered before they can be used with Alexa.

CANCEL   DISCOVER DEVICES

FIGURE 9-8:
When a skill is enabled, tap Discover Devices to see what's available.

10. Tap Done.

With a manufacturer's Alexa skill enabled, you can discover new devices by following Steps 1 through 5 and then tapping Discover Devices, or you can ask Alexa to run the following voice request:

"Discover my devices."

With your smart-home devices plugged in, turned on, and connected to Alexa, you're ready to reap the harvest of all that labor: controlling those devices through Alexa using voice requests. Don't let all that power go to your head!

FIGURE 9-9:
You see a screen similar to this one if the Alexa app discovers any devices using the manufacturer's Alexa skill.

REMEMBER

In the sections that follow, I outline the Alexa voice requests that are generally available for each type of device. Keep in mind, however, that the ways you can control a smart-home device through Alexa are almost always only a subset of what you can do using the manufacturer's app. With a smart plug, for example, Alexa can only turn the device on or off, but the manufacturer's app will usually let you schedule on/off times, turn the plug off automatically after a set time, and more.

Controlling a smart-home device

Before getting to the meat of this section, you should know that there are actually three methods you can use to control a smart-home device:

» **Voice requests:** This is how you'll operate most of your smart-home devices. The rest of this section takes you through the most common voice requests for a selection of smart-home devices.

» **Alexa app:** If you have your Alexa device microphone turned off, you can still use the Alexa app to control your smart-home devices. Tap Devices, tap the device type (or All Devices), and then tap the device you want to mess with. The screen that appears contains the controls you can use. For example, Figure 9-10 shows the device screen for a smart lightbulb, which includes two controls: a button for turning the device on and off and a slider for setting the brightness.

DESK LAMP

Light is **off**
Brightness set to **75%**

FIGURE 9-10:
The device control screen for a smart lightbulb.

>> **Alexa device with a screen:** Swipe down from the top of the screen to open the status bar, and then tap the icon for the device type (such as a bulb icon for your smart lights, plugs, and switches). Note, too, that after you issue a smart-home-device-related request to an Alexa device with a screen, you see some device controls on the screen for a few seconds.

Turning smart plugs on and off

If you're curious about smart-home technology, but you don't want to spend a ton of money or time, a smart outlet — most often called a *smart plug* — is the way to go. A smart plug is an electrical outlet that you can control with voice requests. The smart outlet plugs into a regular electrical outlet for power, and then you plug a non-smart device — such as a lamp or coffeemaker — into the smart outlet. *Voilà!* You now have voice control over the dumb device.

TIP

Do you have a bunch of nearby dumb devices that you want to control via Alexa? In that case, instead of getting multiple smart plugs, buy a single smart power strip.

Note, however, that "control" here just means turning the device on and off using the following voice requests:

>> "Turn [*device name*] on."

>> "Turn [*device name*] off."

Replace *device name* with the name you gave to the smart plug using either the manufacturer's app or the Alexa app.

Working with smart lights

Another easy and relatively inexpensive way to get your smart-home feet wet is with a smart lightbulb or two. You can buy a smart lightbulb for less than $20, and installing it is as easy as changing any regular lightbulb. You can also get smart lightbulbs that change brightness without a separate dimmer switch and that can display different colors.

What if you have a large collection of lights in, say, your kitchen or living room? Swapping out all those dumb bulbs for smart versions would cost a fortune, so a better choice is a smart light switch that you can turn on and off with Alexa. For more control, you can get a smart dimmer switch that enables you to control the brightness with voice requests.

WARNING

Although a smart lightbulb is easy to install, a smart light switch is another matter because it needs to be wired to your home's electrical system. Unless you really know what you're doing, hire an electrician to do the installation for you.

Here are the voice requests to use to turn a smart lightbulb or light switch on or off:

>> "Turn [*device name*] on."

>> "Turn [*device name*] off."

For dimmable smart lights (or smart dimmer switches), use any of the following voice requests:

>> "Brighten [*device name*]."

>> "Dim [*device name*]."

>> "Set [*device name*] brightness to [*number*] percent."

For smart lights that support different colors, use these voice requests:

>> "Set [*device name*] to warm white."

>> "Set [*device name*] to cool white."

>> "Set [*device name*] to [*color*]." (For example, "Set Chill Room to blue.")

4
The Part of Tens

Learn how to use Fire TV to finally get rid of your expensive cable account.

Take a look at some common problems with Fire TV and learn how to solve them.

Ensure that your Fire TV device and your Amazon account are private and secure.

Chapter 10

Cutting the Cord: Ten Steps to Going Cable-Free

Did you know that a *cord-hater* is someone who dislikes paying for cable TV? That person may then become a *cord-shaver*, who takes steps to reduce her cable TV bill. Either way, that person can't ever call herself a *cord-never*, which refers to those enviable (and probably very young) folks among us who've never had a cable TV subscription.

However, a cord-hater can definitely become a *cord-cutter*, meaning someone who severs her relationship with her cable company and finds alternatives to cable elsewhere. That elsewhere may well be Fire TV because it offers tons of alternatives to cable, from Amazon Prime Video to Netflix to YouTube.

If you find yourself with an ever-increasing amount of steam coming out of your ears every time you pay your cable bill, then you may be ready to take the cord-cutting plunge. If so, this chapter offers you a ten-step program for using Fire TV to go cable-free. If that sounds like bliss to you, then read on, cord-cutter, read on.

Step 1: Decide If You Really Want to Cut the Cord

I show in this chapter that although cutting yourself free of cable TV isn't complicated, it does require a nontrivial amount of dedication, a kind of stick-to-it-iveness, if you will. So there's no point starting down the cord-cutting road unless you're absotively, posilutely sure that you want to live cable-free.

To help you get to that level of certainty (if you're not there already, that is; if you are, feel free to skip ahead to the next section), here's a list of reasons why you may *not* want to cut cable from your life:

>> **You may not save as much money as you think.** The main reason to get the cable company out of your life is to reduce that highway-robbery-like payment we all seem to fork over every month. However, although there are some free and inexpensive streaming options out there, to get the shows you and your family really want to watch, you usually have to pay for the privilege (particularly for popular services such as HBO and live sports).

>> **You may not be able to go bundle-free.** One of the things that really irks most cable customers is wanting to watch a particular channel, but having to buy a "bundle" of five or six channels to get the one you want. Inevitably, those extra channels go unwatched, so it feels like you're paying the total price of the bundle just for a single channel. Grr. Alas, I wish I could tell you that such bundles go out of your life when you cut the cord, but many streaming services pull the same stunt by hiding premium channels (such as HBO) with other offerings.

REMEMBER

Some streaming services now offer what are known in the trade as *skinny bundles,* which include just a few channels, making them cheaper than the "fat" bundles offered by cable companies. The cord-cutter's nirvana is *à la carte service,* where you get to select (and pay for) just the channels you want. Unfortunately, very few streaming services offer an à la carte option. One à la carte option you can check out is Amazon Prime Video Channels, which offers monthly subscriptions to individual channels such as HBO and Starz.

>> **You may not be able to go ad-free.** Another thing that really sticks in the craw of most cable users is that we pay a queen's ransom each month and we *still* have to watch way too many obnoxious ads. You may believe that banishing cable from your life comes with the bonus of an ad-free TV experience. Although it's true that most streaming services don't show ads, be warned that some do, especially free services. And just because a service is ad-free now, doesn't mean it always will be. (Netflix, for example, has tested showing ads between episodes of TV shows.)

>> **Your TV life will become *more* complicated, not less.** About the only good thing you can say about cable service is that it's simple: You pay a single (exorbitant) fee each month, and your channels and apps come to you in a single package. That simplicity goes out the window when you cut the cord because now you have to set up separate accounts for each streaming service, pay separate bills, deal with multiple apps, learn a new interface for each app, and somehow remember which service provides which shows. Fire TV helps immensely by bringing everything together under one roof, but cable-free will always be more complicated than cable.

REMEMBER

Unfortunately, the complication of going without cable is getting worse, not better. It seems that practically every week some big-time media company announces that it, too, is jumping on the streaming bandwagon with yet another service that wants to drain $5 or $10 (or more) from your bank account each month. It's madness, but that's life in the cable-free lane.

>> **Your Internet usage will skyrocket — and so may your bill.** By definition, streaming media services are delivered to you through your Wi-Fi network's Internet connection. And, also by definition, video streams such as TV shows and (especially) movies send massive amounts of data to your network. This is meaningless if you have an unlimited data usage plan with your Internet service provider (ISP), but if your ISP caps your monthly usage, going over that ceiling because you binge-watched *Breaking Bad* may cost you big bucks.

>> **Actually, your Internet bill will probably go up no matter what.** If you're like most people, you get your cable TV as part of a bundle that also includes Internet service. If you cut out the cable TV part so that you're paying for just Internet service, you lose the bundle discount and your Internet bill goes up.

>> **The video quality may (not to put too fine a point on it) suck.** Streaming video requires a fast Internet connection, although the required speed — measured in megabits per second, or Mbps — depends on the quality of the video stream:

Video Quality	Bare Minimum Speed	Acceptable Speed
Standard Definition (SD)	1 Mbps	3 Mbps
High Definition (HD)	5 Mbps	8 Mbps
4K Ultra-High Definition (UHD)	18 Mbps	25 Mbps

If your Internet connection's download speed can't match these speeds, then you either need to try a lower-quality stream or put up with pauses while the content buffers (see Chapter 1), playback stutters, or poor video quality.

>> **You may have to wait a while before you can watch "new" TV shows.** Not surprisingly, cable companies get first crack at new cable shows. Those shows may eventually get streamed on services such as Netflix, but it could be as much as a year later.

>> **Or you may *never* see "new" TV shows.** Another aspect of the increasing complexity of the streaming universe is that many services are now spending zillions of dollars to create their own content. Think of *The Handmaid's Tale* on Hulu, *House of Cards* on Netflix, and *The Marvelous Mrs. Maisel* on Amazon Prime Video. It's possible some of this content may be seen on other services, but the majority of it will stay within the "walled garden" of the original streaming service, meaning that if you don't have a subscription for that service, then you don't get to see the show.

Okay, have I scared you away from cutting the cord? If so, I understand. If not, good for you! Now you're ready to take the next step.

Step 2: Make a List of Your "Must-See" Shows (Or Not)

Do you want to know the secret of people who have both cut the cord *and* remained happy that they did? It's nothing esoteric or complex. In fact, I can tell you everything you need to know in just a few simple words:

> They didn't try to clone their cable setup.

Yep, that's it. Would-be cord-cutters often think they need to figure out a way to replicate their cable configuration using streaming services. And, yes, you can absolutely do that, but be forewarned that you'll almost certainly end up paying just as much per month (and possibly more) as you currently do for cable, with the added inconvenience of juggling multiple services. Forget about it.

Instead, successful and happy cord-cutters travel one or the other of the following paths to post-cable satisfaction:

>> Make a list (ideally, a very short one) of the shows or other content (such as live sports) that you can't live without. Then figure out which streaming services offer those shows.

WARNING

Getting your sports fix without cable is a tricky and usually very expensive proposition. Watching one or two sports without cable usually isn't so bad, but sports junkies who want to watch everything are almost always better off sticking with a sports bundle offered by the cable company.

» Pick one or two streaming services that offer lots of content, knowing that, unless you're extremely fussy about your entertainment, you're always going to find something interesting or fun to watch. For example, a Netflix subscription combined with Amazon Prime Video (assuming you're a Prime member) offers more movies and TV shows than you'll ever be able to watch, all for just a small monthly amount.

Augment these streaming services with over-the-air live TV (see Step 4) for some news, sports, and perhaps a few primetime shows, and you've got the makings of a decent cable-free setup.

Step 3: Figure Out What Equipment You Need

Assuming you already have either a Fire TV Edition Smart TV or a separate Fire TV device that's connected to a television, what other equipment do you need before you cut the cord?

If your current modem/router is a cable company rental, then after you go sans cable, you'll need to replace the device. Here are the main things to look for in a new modem/router:

» Make sure the modem is compatible with your ISP.

» Make sure the modem can support your Internet download speed. For example, a modem that can only support up to 250 Mbps is no good if you're paying for 500 Mbps downloads.

» The faster the Wi-Fi speed the better, because your media streams have to get to your devices via your Wi-Fi network.

» Get a multi-band (for example, dual-band or tri-band) router, which enables the router to prioritize streaming content over other types of network traffic.

If you also want to view over-the-air live TV, then you'll need an HDTV antenna. See Step 4 for some tips about buying an antenna that's right for your needs.

Step 4: Check What's Available Over-the-Air

Perhaps the simplest — and certainly the cheapest — way to do live TV is to purchase an HDTV antenna and connect it to your TV. This works because TV stations around the country — including network affiliates and independent stations — broadcast their live TV signals into the ether. These so-called *over-the-air* (OTA) signals can be picked up by an HDTV antenna.

What's the cost? Well, aside from the cost of the antenna itself, OTA signals are absolutely free! What's the catch? OTA signals can travel only a finite distance, so getting a strong signal depends on several factors:

>> **The strength of the original broadcast signal:** Some stations (especially those associated with big-time networks) emit powerful signals that can travel long distances, but most signals don't travel very far.

>> **Your location:** If you live in a big city, you likely have quite a few stations broadcasting nearby; if you live in the country, most broadcasters are likely to be quite far away.

>> **The range of your antenna:** Cheap indoor antennas have ranges of just a few miles, while expensive outdoor antennas can pick up signals 60 or 70 miles away.

>> **Large obstacles in the signal path:** If tall buildings, big hills, or other large objects are between your antenna and a signal, the signal will be weaker than it would if no such objects were in the way.

>> **The direction of your antenna:** Generally speaking, your antenna will be more likely to pick up an OTA signal if the antenna is pointed at the station broadcasting the signal.

>> **The phase of the moon:** I jest, of course, but with the often mercurial nature of OTA signals, it sometimes feels like the moon or some other intangible force is messing with your TV-watching pleasure.

Given the large number of signal-strength factors in the preceding list, you may be tempted to go out and buy the most powerful antenna available. Sure, you *could* do that, but you don't want to end up with too much antenna if you don't need it. How can you know? The easiest way is to crank up your favorite web browser, send it to the TV Fool website (www.tvfool.com), and then click the Signal Analysis tab. This opens TV Fool's TV Signal Locator tool, which uses your address (or your

GPS coordinates) to give you a report of the broadcast TV signals that are available to you, as well as the relative strength of those signals. To use TV Fool, follow these steps:

1. **Select the Address radio button and then enter your location: street address, city, state or province, and zip or postal code.**

If you don't want to provide your street address, you can select the Coordinates radio button and then enter your location's latitude and longitude.

TIP

To get your location's latitude and longitude, use Google Maps (https://maps.google.com) to search for your location. Right-click the pin that appears on your location and then click What's Here? in the shortcut menu. Google Maps displays a card at the bottom of the screen that includes, in order, your latitude and longitude.

2. **Click Find Local Channels.**

TV Fool returns after a few seconds with the results, which will resemble those shown in Figure 10-1. The radar chart on the left shows the direction for each signal, while the chart on the right shows the signals you can pick up from your location.

All Channels

TrueNorth

Search Criteria

Address: exact
Hoboken, NJ
Postal code 07030

db datecode
201801291705

www.tvfool.com

Callsign	Real	(Virt)	Netwk	NM(dB)	Pwr(dBm)	Path	miles	True	(Magn)
WARC-TV	7		ABC	79.2	-11.6	LOS	1.9	154°	(167°)
W1LP	3	(33.1)		78.9	-12.0	LOS	2.6	60°	(72°)
WXTV-DT	40	(41.1)	Uni	76.4	-14.5	LOS	2.4	71°	(84°)
WCBS-DT	33	(2.1)	CBS	75.9	-14.9	LOS	2.4	71°	(84°)
WNBC	28	(4.1)	NBC	74.9	-15.9	LOS	2.4	71°	(84°)
WNYW-DT	44	(5.1)	FOX	74.4	-16.4	LOS	2.4	71°	(84°)
WPXN-TV	31	(31.1)	ION	73.9	-16.9	LOS	2.4	71°	(84°)
WNYE-DT	24	(25.1)	Ind	73.4	-17.5	LOS	2.6	60°	(72°)
WWOR-TV	38	(9.1)	MyN	73.3	-17.6	LOS	2.4	71°	(84°)
WNJU	36	(47.1)	TLL	73.6	-19.2	LOS	2.4	71°	(84°)
WFUT-DT	26		Tel	70.5	-20.3	LOS		71°	(84°)
WPIX	11	(11.1)	CW	69.6	-21.2	LOS	1.9	113°	(167°)
WPXO-LD	4	(34.1)		65.1	-25.7	LOS	2.4	71°	(84°)
WKOB-LD	2	(42.1)		63.8	-27.1	LOS	3.4	72°	(85°)
WDVB-CD	23	(23.1)		58.3	-32.6	LOS	2.4	71°	(84°)
WBQM-LD	50			57.4	-33.4	LOS	3.4	72°	(85°)
WASA-LD	25	(24.1)		52.0	-38.8	LOS	2.6	60°	(72°)
WE8R-CD	49			49.8	-41.0	LOS	2.4	71°	(84°)
WMBC-TV	18	(63.1)	Ind	42.9	-47.9	1Edge	12.6	314°	(327°)
WMBC-TV	18		Ind	42.3	-48.6	LOS	2.4	71°	(84°)
WMUN-CD	45			41.3	-49.6	LOS	3.4	72°	(85°)
WNJN-DT	51	(50.1)	PBS	38.5	-52.4	1Edge	12.6	314°	(327°)
WNXY-LD	10			38.3	-52.5	LOS	4.5	81°	(94°)
W41DO-D	41	(60.1)	HSN	38.2	-52.7	LOS	3.6	62°	(73°)
WNYJ-TV	29	(66.1)		38.2	-52.7	2Edge	12.4	287°	(299°)
WMBC-TV	18		Ind	36.8	-54.0	1Edge	12.6	314°	(327°)
WNJB	8	(58.1)	PBS	34.4	-56.5	2Edge	26.2	252°	(265°)
WMBQ-CD	46			26.2	-64.6	LOS	3.4	72°	(84°)
WNVN-LD	39	(39.1)		18.7	-72.1	LOS	4.5	81°	(94°)
W20EF-D	20			18.7	-72.1	LOS	3.4	72°	(85°)
WXNY-LD	22			14.2	-76.7	LOS	4.5	81°	(94°)
WNYX-LD	5			12.8	-78.0	LOS	4.5	81°	(94°)
WYXN-LD	9			11.8	-79.1	LOS	4.5	81°	(94°)
WNYZ-LP	6			10.5	-68.3	LOS	4.5	71°	(84°)
W2RES-D	28	(49.1)		9.9	-80.9	2Edge	12.4	287°	(299°)
WNYZ-LP	6	(6.1)		9.1	-81.8	LOS	4.5	81°	(94°)
WTBY-TV	27	(54.1)	Ind	8.5	-82.3	2Edge	52.0	5°	(17°)
WRNN-DT	48	(48.1)	Ind	5.9	-85.0	2Edge	52.0	5°	(17°)
WPVI-TV	6	(6.1)	ABC	5.0	-85.9	2Edge	80.2	234°	(246°)
KJWP	2			4.4	-86.4	2Edge	79.8	233°	(246°)
WNJT-DT	43	(52.1)	PBS	3.0	-87.8	2Edge	46.7	228°	(241°)
WACP	4			1.1	-89.7	2Edge	81.4	212°	(225°)
W43CH-D	43	(43.1)		-3.7	-94.6	2Edge	46.7	228°	(241°)
WLNY-DT	47	(55.1)	Ind	-6.6	-97.4	2Edge	59.4	75°	(87°)
WNMF-LD	17			-7.5	-98.4	2Edge	25.5	278°	(290°)
W25FA-D	25	(24.1)		-7.6	-98.5	2Edge	31.3	303°	(315°)
WPHL-DT	17	(17.1)	MyN	-8.0	-98.9	2Edge	79.9	233°	(246°)
KYW-TV	26	(3.1)	CBS	-8.2	-99.0	2Edge	80.0	234°	(246°)
WTNH-DT	10	(8.1)	ABC	-8.2	-99.0	2Edge	73.5	59°	(72°)
W09CZ-D	9			-8.8	-99.6	1Edge	19.4		
WBPH-TV	9	(60.1)	Ind	-10.3	-101.1	2Edge	74.8	261°	(274°)

FIGURE 10-1: A typical broadcast channel report from TV Fool.

The signals are listed in descending order of strength and are color-coded as follows:

- **Green:** These signals are strong enough that you can probably pick them up using a simple indoor antenna.

- **Yellow:** These signals are less strong, so you may need a larger attic-mounted antenna to pick them up.

- **Red:** These signals are relatively weak, so you probably need a roof-mounted antenna to pick them up.

- **Gray:** These signals are probably too weak to pick up with even the most powerful antenna.

Step 5: Make a Streaming Budget

In some ways, making a budget for your streaming activities is the most important step because, after all, you're probably cutting the cord to save money. Without a budget, it's easy to add one service for $10 a month, another for $15, yet another for a mere $5.99, and before you know it you're forking over far more each month than you ever did for cable.

You've already made a list of your "must-see" shows (see Step 2), so now you need a second list that includes the monthly cost for each streaming service you want to use:

» General streaming services such as Amazon Prime Video and Netflix

» Premium channel services, such as HBO and Starz

» Sports services, either channels such as ESPN or specific sports services such as MLB.TV and NFL Live

Don't forget to include the extra amount you'll most likely be paying each month for Internet access:

» If your Internet connection is currently bundled with your cable package, your Internet cost will rise without that bundle.

» If you already get Internet access from a provider other than your cable company, you may have to pay more to increase your monthly data usage to allow for your new streaming lifestyle.

CABLE-REPLACEMENT SERVICES

I realize that saving money isn't the only reason to say "so long" to your cable overlords. It may be more important to your future happiness to simply get the cable company out of your life, period. If that's the case, and you don't care so much about saving money, then you should look into the so-called *cable-replacement services*. These are streaming services that offer TV show and movie bundles just like cable, but without actually being cable.

Here are some popular cable-replacement services in the United States to check out:

- Hulu + Live TV (www.hulu.com/live-tv)
- Sling TV (www.sling.com)
- YouTube TV (https://tv.youtube.com)

In Canada, check out Crave TV (www.crave.ca).

Add it all up, and as long as you're coming in under (hopefully, well under) your current cable cost, you're ready to move on.

Step 6: Subscribe to Streaming Services

When you know which streaming services you want to use, fire up your trusty web browser and visit the website of each service to sign up for an account. This sounds like scary commitment time, but happily most streaming services offer a free trial period, typically a month. That gives you plenty of time to try out the service free to make sure it's what you want.

Some services offer a discount if you pay for a longer-term subscription upfront. For example, a service that's $9.99 per month may offer a 12-month subscription for $99. That's like getting two months free, which is fine if you're certain you'll use the service for the next year. Because you're just starting out, you really don't have any way of knowing whether you'll stick with any new streaming service for the long haul, so at least at first you're better off signing up for month-to-month subscriptions. Then down the road, if you really like what a service has to offer, you can extend your subscription to get the discount.

If a streaming service has just one show you really want to watch, it doesn't make economic sense to pay even a small monthly fee for years on end just to watch a single show. Instead, wait until the service offers the full season of the show. Then sign up for a subscription (ideally a free trial), binge-watch the entire season, and then cancel the subscription. Worst-case scenario: You see your show's entire season for the cost of a month's service.

Step 7: Set Up Fire TV Recast

Unless you sign up with a cable-replacement service, one feature you may lose when you cancel your cable account is recording live TV shows. Most cable-replacement services include a cloud-based digital video recorder (DVR), but if you're getting your live TV over-the-air through an HDTV antenna, you can't record anything.

Or, I should say, you can't record anything unless you add Amazon's Fire TV Recast to your entertainment system. As I describe in Chapter 7, Fire TV Recast is a DVR that enables you to watch and record over-the-air shows picked up by an HDTV antenna. It's just the ticket if you hate having to watch live TV when it's actually live, so see Chapter 7 for all the details.

Step 8: Put It All Together with Fire TV

As I mention in Chapter 1, today's streaming-media environment is becoming super-complicated, with new services coming online at an alarming rate. It's a rare cord-cutter who sticks with a single service such as Amazon Prime Video or Netflix. Instead, we all have multiple streaming services, but that breeds even more complications because we now have to navigate multiple websites, multiple logins, multiple account renewals, and so on.

One of the main themes of this book is that you can greatly reduce the complexity of today's media-streaming environment by bringing all your streaming services under the big tent of Fire TV. Sure, Fire TV is optimized for Amazon Prime Video (no surprise there), but having the likes of Netflix and YouTube (among many others) right there in the Fire TV interface is very convenient.

Step 9: Do a Trial Run

Soon you'll pull the plug on your cable account officially. I know, I know, you can't wait. But before you make things official, a good short-term option is to pull the plug *unofficially.* That is, keep your cable service for now, but just disconnect the physical cable from your TV. That way, you can live "cable-free" for a week or three to see if you like it.

Sure, technically, you don't have to physically unplug your cable connection. You can leave the cable where it is and just keep your television tuned to the Fire TV input source (switching as needed to the antenna input source, if you're using an HDTV antenna for live TV and you don't have a Fire TV Edition Smart TV). However, I think it's better to yank that cable away from your TV because otherwise it's just too easy and too tempting to switch over to the cable input source to catch the game or some other content that's missing from the cable-free side of things.

During your trial run, feel free to sign up for tons of streaming services. In an ideal world, you'd only sign up for services that offer the first month (or whatever) free so that you're not "double-dipping" by paying for both cable and non-cable services at the same time.

TIP

One hazard with trying out lots of streaming services is that you may lose track of one or two and end up paying for services you don't want when the trial periods expire. To prevent such unwanted charges, for each service you sign up for, immediately cancel your account. Almost all streaming services will still give you full access until the end of the month, so you can continue your trial with no worries. If you decide you want to keep the service, you can reinstate your account.

Step 10: Say Goodbye to Your Cable Company

TIP

At long last, the day has come to (if I may paraphrase a famous line from the movie *Apocalypse Now*) terminate your cable account with extreme prejudice. You may hope at this point that a simple phone call will suffice, but is anything ever simple when it comes to the cable company? Don't be silly. Here are a few tips to keep in mind to help the process go, well, if not smoothly, at least somewhat bearably:

>> **Before calling, check your contract to see if the cable company has the right to charge you a cancellation fee.** Note that you may be able to get the

cable company to reduce or waive that fee if you tell the representative that you'll be staying with the company for Internet access.

>> **Before calling, sign in to your account and check your Internet data usage history.** If you find that you're consistently near the current usage cap, know that you'll have to increase that ceiling if you plan on staying with the cable company for Internet access.

>> **Remind yourself that "conversations" with cable company employees are always frustrating, at best, and downright rage-inducing, at worst.** This phone call will be no different, believe me.

WARNING

Actually, this phone call will almost certainly be *worse* than usual because you'll have to deal with a cable company denizen called the *retention agent.* As the name implies, that employee's job is to retain you as a customer, and the lengths most retention agents usually go to keep you as a customer can be quite frustrating. My only advice is stick to your guns and don't back down.

>> **Have all your account information (particularly your account number) handy.**

>> **Have a pen and paper within reach to take notes during the phone call.**

>> **Learn (and write down) the name of the rep who answers your call.**

>> **Be prepared for a very long wait on hold.** This is a tactic — the cable company hopes you'll eventually hang up. Have a coffee and/or crossword puzzle handy to keep you occupied.

>> **If you plan on sticking with the cable company for Internet access, make sure you find out what an Internet-only account will cost you each month.** Don't be afraid to ask if the company currently has any deals for Internet access.

>> **Before hanging up, get the agent to confirm that your cable service is cancelled and that there will be no hidden or extra fees charged to you.**

>> **If you have equipment to return (such as a set-top box and/or a modem), be sure to take it (or send it) back as soon as possible.** Don't give the cable company any excuse to ding you with "late fees" or other extra charges.

Chapter **11**

Ten Things That Can Go Wrong

F ire TV devices such as the Fire TV Stick, Fire TV Stick 4K, and Fire TV Cube are extremely simple, Spartan even, when viewed from the outside. They're devoid of moving parts; the Fire TV Stick and Fire TV Stick 4K have but a single port for the power supply; and the Fire TV Cube has just a few ports in the back. But don't let that Zen-inducing outer shell fool you — inside every Fire TV device is a complex and sophisticated array of electronics. The good news is that these intricate innards enable Fire TV to perform its streaming magic; the bad news is that, when it comes to electronic devices, "complex and sophisticated" almost always leads to some kind of trouble. That's just a fact of modern life, although it doesn't guarantee that your Fire TV device will one day bite the digital dust. In fact, that's very unlikely because Fire TV devices are known to be remarkably robust and nearly error-free.

Did you notice that hedge-word *nearly* in that last sentence? Alas, yes, sometimes even Fire TV devices behave strangely. In this chapter, I fill you in on ten of the most common problems related to Fire TV software and hardware, and show you how to solve every one of them.

Troubleshooting General Problems

TIP

Before getting to the specific problems and their solutions, I want to take you through a few very basic troubleshooting steps. Many problems, particularly problems related to your Fire TV device, can be solved by doing the following three things (each of which I explain in more detail in the sections that follow):

» Restart your Fire TV device.

» Update your Fire TV device's system software.

» Reset your Fire TV device to its factory default settings.

REMEMBER

Try restarting your Fire TV device to see if it solves your problem. If not, move on to updating the software and see if that helps. If there's still no joy, only then should you try resetting your Fire TV device to its factory default settings.

Restarting your Fire TV device

If your Fire TV is having trouble playing media, connecting to Wi-Fi, pairing with a Bluetooth device, or doing any of its normal duties, by far the most common solution is to shut down the device and then restart it. By rebooting the device, you reload the system, which is often enough to solve many problems.

There are three ways to restart either a TV that has a Fire TV device attached or a Fire TV Edition Smart TV:

» If you still have access to the Fire TV interface, choose Settings, choose either My Fire TV (for Fire TV Stick or Fire TV Cube) or Device & Software (for Fire TV Edition), and then choose Restart.

» On the Fire TV remote, press and hold both the Select button and the Play button for about five seconds until Fire TV restarts.

» For a Fire TV device or a Fire TV Edition Smart TV, unplug the device's power cord, and then plug it back in.

WARNING

You may be tempted to just plug the Fire TV device back in again right away, but hold on a second. The Fire TV device has internal electronic components that take some time to completely discharge. To ensure you get a proper restart, wait at least three seconds before reconnecting the Fire TV power supply.

Checking your Fire TV device for software updates

Your Fire TV device uses internal software — called Fire OS (*OS* is short for *operating system*) — to perform all sorts of tasks, including connecting to your Wi-Fi network, handling media playback, and saving your settings. If your Fire TV is acting weird, and restarting the device doesn't help, you can often un-weird the device by updating Fire OS. Sometimes installing a new version of the operating system is all you need to make your problem go away. In other cases, updating the system may fix a software glitch that was causing your problem.

Here are the steps to follow to check for and install Fire OS updates:

1. **Choose Settings.**

2. **Choose either My Fire TV (for Fire TV Stick or Fire TV Cube) or Device & Software (for Fire TV Edition).**

3. **Choose About.**

4. **Choose Check for Updates.**

   When you highlight the Check for Updates command, the right side of the screen shows the current version of Fire OS, as well as the last date Fire TV checked for updates, as shown in Figure 11-1.

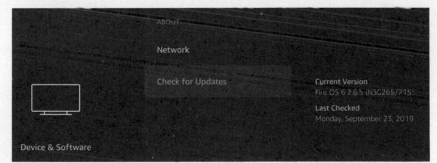

FIGURE 11-1: Highlight Check for Updates to see your Fire OS version and last checked date.

Fire TV checks to see if an updated version of Fire OS is available. If an update is waiting, Fire TV downloads the update and then displays the Install Update command, as shown in Figure 11-2.

Fire TV installs this update automatically the next time it determines that you're not using your TV.

FIGURE 11-2:
The Install
Update command
appears when
Fire TV has
downloaded an
update to Fire OS.

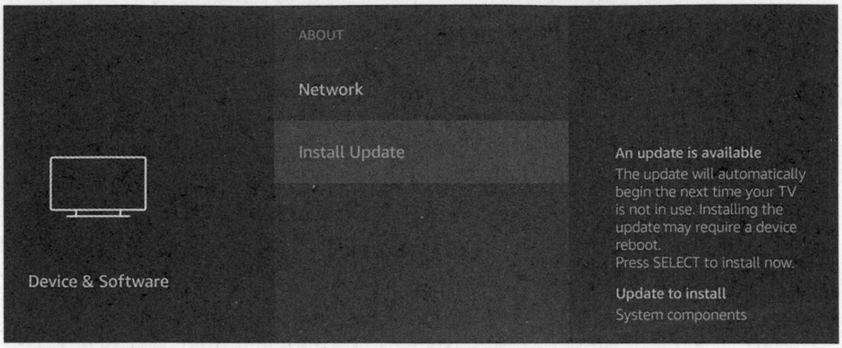

5. **If you'd rather install the update right away, choose Install Update.**

   Fire TV installs the update. During this process, leave your device on and don't press any buttons on the remote.

Resetting your Fire TV device

If your problem is particularly ornery, restarting or updating the device won't solve it. In that case, you need to take the relatively drastic step of resetting your Fire TV device. I describe this step as "drastic" because it means you have to go through the setup process all over again, so only head down this road if restarting and updating your device don't solve the problem.

How you perform the reset depends on your device:

>> **Fire TV Stick or Fire TV Cube:** Choose Settings ⇨ My Fire TV ⇨ Reset to Factory Defaults.

>> **Fire TV Edition:** Choose Settings ⇨ Device & Software ⇨ Reset to Factory Defaults.

You Can't Connect to Your Wi-Fi Network

Wireless networking adds a whole new set of potential snags to your troubleshooting chores because of problems such as interference and device ranges. Here's a list of a few troubleshooting items that you should check to solve any wireless connectivity problems you're having with your Fire TV device:

>> **Restart your devices.** Reset your hardware by performing the following tasks, in order:

1. Turn off your modem.

2. Turn off your Wi-Fi router.

3. After a few seconds, turn the modem back on and wait until the modem reconnects to the Internet, which may take a few minutes.

4. Turn on your Wi-Fi router.

REMEMBER

Many Wi-Fi devices these days are all-in-one gadgets that combine both a Wi-Fi router and a modem for Internet access. If that's what you have, instead of performing Steps 1 through 4, you can just turn off the Wi-Fi device, wait a bit, turn the device back on, and then wait for the device to connect to your Internet service provider (ISP).

5. Restart Fire TV as I describe earlier in the "Restarting your Fire TV device" section.

>> **Look for interference.** Devices such as baby monitors and cordless phones that use the 2.4 GHz radio frequency (RF) band can wreak havoc with wireless signals. Try either moving or turning off such devices if they're near your Fire TV device or Wi-Fi device.

WARNING

Keep your Fire TV device and Wi-Fi router well away from microwave ovens, which can jam wireless signals.

TIP

Many wireless routers enable you to set up a separate Wi-Fi network on the 5 GHz RF band, which isn't used by most household gadgets, so it has less interference. Check your router manual to see if it supports 5 GHz networks.

>> **Check your password.** Make sure you're using the correct password to access your Wi-Fi network.

>> **Check your range.** Your Fire TV device may be too far away from the Wi-Fi router. You usually can't get much farther than about 230 feet away from most modern Wi-Fi devices before the signal begins to degrade (that range drops to about 115 feet for older Wi-Fi devices). Either move the Fire TV device closer to the Wi-Fi router or, if it has one, turn on the router's range booster. You could also install a wireless range extender.

>> **Check your password.** Make sure you're using the correct password to access your Wi-Fi network.

>> **Update the wireless router firmware.** The wireless router firmware is the internal program that the router uses to perform its various chores. Wireless router manufacturers frequently update their firmware to fix bugs, so you should see if an updated version of the firmware is available. See your device documentation to find out how this works.

>> **Update and optionally reset your Fire TV device.** Make sure your Fire TV device is up to date (see "Checking your Fire TV for software updates," earlier in this chapter) and, if you still can't connect to Wi-Fi, reset your Fire TV device (see "Resetting your Fire TV device," earlier).

>> **Reset the Wi-Fi device.** As a last resort, reset the Wi-Fi router to its default factory settings (see the device documentation to find out how to do this). Note that if you do this, you need to set up your network again from scratch.

You're Having Trouble Streaming Media

Streaming media on Fire TV works well most of the time, but problems can arise, particularly the following:

>> The media never starts.

>> The media takes a long time to start.

>> The media plays intermittently.

>> The media stops playing and never resumes.

It's maddening, for sure, but most of the time you can fix the problem. I say "most of the time" because there are a couple of situations where media streaming just doesn't work well:

>> **When you have a slow Internet connection speed:** Media files are usually quite large, so for these files to play properly you need a reasonably fast Internet connection. Amazon recommends at least a half a megabit per second (0.5 Mbps), but realistically you should probably have a connection that offers at least 8 Mbps download speeds for HD content (see Chapter 10 for more detail on this).

>> **When you have an intermittent Internet connection:** If you live in an area with spotty Internet service, that now-you-see-it-now-you-don't Internet connection makes streaming media impossible.

What if you have a zippy Internet connection and strong service all the time? First, congratulations! Second, there are a few things you can try to get media streaming to work better (or at all). Try these troubleshooting ideas in the following order:

>> Restart your Wi-Fi router and Fire TV device using the steps I outline in the preceding section.

» If you have other devices accessing your Wi-Fi network, shut down any devices you're not using.

» Move your Fire TV device closer to your Wi-Fi router. Your Fire TV device must be within 230 feet of the router (115 feet for older routers), but the closer the two devices, the stronger the Wi-Fi signal.

» Make sure your Wi-Fi router isn't situated near devices that can cause interference, such as microwave ovens and baby monitors.

» Make sure your Fire TV device isn't close to a wall or a metal object.

» If your Fire TV device is sitting on a low shelf or even on the floor, move it to a higher location.

» If your Fire TV device is inside a cabinet or similar enclosure, take it out.

» If you're having streaming problems with a particular app, run through the following steps to clear the app's application data:

WARNING

An app's data includes your app account info and any settings you've configured for the app. If you clear the app's data, it means you have to enter your login info all over again and reconfigure your settings, so be sure you want to perform this operation before proceeding.

1. **On your Fire TV device, choose Settings ⇨ Applications ⇨ Manage Installed Applications.**

2. **Choose the app you're having trouble with.**

3. **Shut down the app by choosing the Force Stop command, shown in Figure 11-3.**

4. **Clear the app's data by choosing Clear Data ⇨ Clear Data command.**

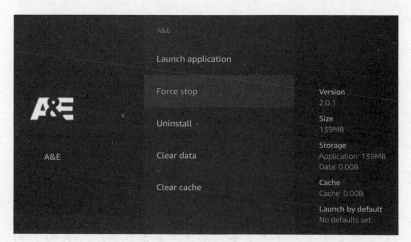

FIGURE 11-3: In an app's settings, choose Force Stop to close the app.

The Fire TV Mobile App Can't Pair with Your Fire TV Device

After you have your Fire TV device set up, the Fire TV mobile app will automatically locate and pair with the device. This enables you to use the app as a remote control for the Fire TV device (and, if you have Fire TV Recast, it also enables you to use the app to watch and record over-the-air TV programs). It's all very convenient, but that convenience goes out the window if the Fire TV mobile app can't locate or pair with your Fire TV device. If that happens, optionally curse your rotten luck and the try these troubleshooting techniques:

>> Make sure your Fire TV device and the device that's running the Fire TV mobile app are connected to the same network.

>> Check to see if another device on your network is currently mirroring the display of your Fire TV device (I show you how to mirror Fire TV to another device in Chapter 6). If your Fire TV device is currently being mirrored, the Fire TV mobile app won't be able to pair with the Fire TV device, so you need to first stop the mirroring.

>> Restart Fire TV as I describe earlier in the "Restarting your Fire TV device" section.

>> Restart the device on which the Fire TV mobile app is installed.

>> Disconnect a Bluetooth device or Fire TV remote (including any device running the Fire TV mobile app) from your Fire TV device. Fire TV supports up to seven connected devices, so if you're already at that limit, your Fire TV mobile app won't be able to pair with the Fire TV device.

>> (Advanced) Log in to your wireless router and search the advanced settings for Multicast support. If you see an option for toggling Multicast, make sure that setting is On. (See your wireless router manual to learn how to log in and access the router's settings.)

>> (Advanced) Log in to your wireless router and check to see if a static IP address has been assigned to your Fire TV device. If so, configure the router to assign an address to the Fire TV device automatically. (Again, see your router manual to learn how to work with IP addresses.)

Your Fire TV Screen Is Blank

If your Fire TV Edition Smart TV or the TV to which you've connected your Fire TV device shows a blank screen, here are a few things to check out:

- » Make sure the TV is plugged in and turned on.

- » Make sure the TV is set to the correct input:

  - For a TV with a Fire TV device attached, switch to whatever input the Fire TV device is connected to.

  - For a Fire TV Edition Smart TV, make sure the TV is using the Fire TV input (press the Home button on the Fire TV remote to display the Fire TV Home screen).

- » For a Fire TV device connected to the TV via HDMI, disconnect and then reconnect the device.

- » If your Fire TV device is connected to your TV with an HDMI cable or HDMI hub, try replacing the cable and/or the hub.

- » On the Fire TV remote, press and hold both the Up button on the navigation ring and the Rewind button for about ten seconds. Fire TV begins running through the available resolutions for the TV and displays each resolution for ten seconds. If you see the text for a particular resolution, as shown in Figure 11-4, choose Use Current Resolution.

- » If you have a second-generation Fire TV Cube (which supports 4K), make sure you're using a high-speed HDMI cable.

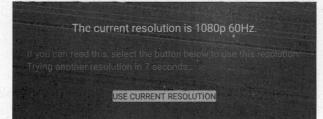

FIGURE 11-4: If your previously blank screen shows the text for a resolution, choose Use Current Resolution to solve the problem.

The current resolution is 1080p 60Hz.

If you can read this, select the button below to use this resolution.
Trying another resolution in 7 seconds...

USE CURRENT RESOLUTION

You Don't Hear Audio During Playback

If the video portion of a stream appears just fine, but you don't hear the audio portion, then some troubleshooting is in order. Here are a few techniques to run through:

1. **Check that the TV audio isn't either muted or set extremely low.**

2. **If your TV is connected to a receiver or similar audio output device, make sure the device is turned on and correctly connected to your TV.**

3. **If you're playing sound through your TV's speaker, make sure the speakers are turned on in the TV's settings.**

    For a Fire TV Edition Smart TV, choose Settings ⇨ Display & Sounds ⇨ Sound Settings, and then set the TV Speakers setting to On.

4. **For a Fire TV device connected to the TV via HDMI, disconnect and then reconnect the device.**

5. **If your Fire TV device is connected to your TV with an HDMI cable or HDMI hub, try replacing the cable and/or the hub.**

6. **For a Fire TV Stick or Cube, try turning off Dolby Digital Plus Output. Choose Settings ⇨ Display & Sounds ⇨ Audio ⇨ Dolby Digital Output, and then choose Dolby Digital Plus Off.**

You're Having Trouble Connecting a Bluetooth Device

Fire TV devices support a wireless technology called Bluetooth, which enables you to make wireless connections to other Bluetooth-friendly devices, such as headsets, speakers, and smartphones. This section provides you with a few common Bluetooth troubleshooting techniques.

You don't see a Bluetooth device

Not surprisingly, you can't make a Bluetooth connection if you can't see the device on the Fire TV device's Add Bluetooth Devices screen (which you display by choosing Settings ⇨ Remotes & Bluetooth Devices ⇨ Other Bluetooth Devices ⇨ Add Bluetooth Devices).

If you don't see a Bluetooth device, try the following:

» Make sure the device is turned on and fully charged.

» Make sure the device is discoverable. Most Bluetooth devices have a switch you can turn on or a button you can press to make them discoverable.

» Make sure the Bluetooth device is well within 33 feet of your Fire TV device, because that's the maximum range for most Bluetooth devices. (Some so-called *Class 1* Bluetooth devices have a maximum range ten times as long.)

>> If possible, reboot the Bluetooth device. If you can't reboot the device, or the reboot doesn't solve the problem, restart your Fire TV device.

>> Check with the Bluetooth device manufacturer to ensure the device is capable of being paired with Fire TV devices. Specifically, you need to find out whether the device supports either of the following Bluetooth profiles:

- Advanced Audio Distribution Profile (A2DP)

- Audio/Video Remote Control Profile (AVRCP)

You can't pair with a Bluetooth device

As a security precaution, many Bluetooth devices need to be paired with another device before the connection is established. You initiate the pairing by tapping the device on the Fire TV device's Add Bluetooth Devices screen (choose Settings ⇨ Remotes & Bluetooth Devices ⇨ Other Bluetooth Devices ⇨ Add Bluetooth Devices).

In some cases, the pairing is accomplished by entering a multidigit passkey — sometimes called a PIN — that you must enter. Fire TV devices don't support Bluetooth PINs, so if your device requires a PIN, you're out of luck.

Otherwise, you may find that even though the device shows up fine on the Add Bluetooth Devices screen, you can't pair it with your Fire TV device.

First, try the solutions in the previous section. If none of those steps does the job, check to see if the Bluetooth device is already paired with another device. Bluetooth stuff can only pair with one device at a time, so before you can pair the device with Fire TV, you need to disconnect the device from its current pairing.

If you still can't get the pairing to work, tell Fire TV to start over by forgetting what it knows about the device:

1. **On your Fire TV device, choose Settings ⇨ Remotes & Bluetooth Devices ⇨ Other Bluetooth Devices.**

2. **Highlight the Bluetooth device name.**

3. **Press the Menu button.**

Fire TV asks you to confirm the unpairing.

4. **Press Select.**

Fire TV removes the device from the Other Bluetooth Devices screen.

5. **Choose Add Bluetooth Devices.**

6. **Press the switch or button that makes the device discoverable.**

7. **When the device reappears on the Add Bluetooth Devices screen, highlight the device and then press Select to pair with it again.**

Your Fire TV Device Is Unresponsive

Perhaps the most teeth-gnashingly frustrating problem you can encounter in technology is when a device — particularly one you paid good money for — just stops working. The device appears to be on, but tapping it, shoving it, gesticulating at it, and yelling at it are all ineffective at making the device respond.

If that happens to your Fire TV device, try the following troubleshooting steps, in order:

1. **Wait a few minutes.**

   Sometimes devices just freeze up temporarily and then right themselves after a short break.

2. **Check your Wi-Fi network to make sure it's working properly and that your device is connected.**

   See "You Can't Connect to Your Wi-Fi Network," earlier in this chapter.

3. **Restart the device.**

   See the "Restarting your Fire TV" section, earlier in this chapter.

4. **Check to see if your device is using the most up-to-date system software.**

   See "Checking your Fire TV device for software updates," earlier in this chapter.

5. **Reset your device.**

   See the "Resetting your Fire TV" section, earlier in this chapter.

You Changed Your Wi-Fi Network Password

Your Fire TV device needs access to your Internet-connected Wi-Fi network to do its thing. When you first set up your Fire TV device, getting the device on your network by entering your network password is one of the first chores. That usually works flawlessly, but what happens if down the road you change your network password? In that case, you need to reconnect to your network by following these steps:

1. **Choose Settings ⇨ Network.**

   You see a list of available networks, with your network at the top of the list.

2. **Highlight your network.**

3. **On the Fire TV remote, press Menu to run the Forget This Network command.**

   Fire TV asks you to confirm that you want to forget the network.

4. **Press the Select button on the remote.**

   The Fire TV device disconnects from the network and returns you to the list of available networks.

5. **Choose your network.**

6. **Enter your new network password and then choose Connect.**

   Fire TV reconnects you to your network.

A Fire TV App Doesn't Work Properly

If a Fire TV app won't start, freezes, or runs erratically, here are some troubleshooting suggestions you can try to resolve the problem:

>> Restart Fire TV (see the "Restarting your Fire TV device" section, earlier in this chapter).

>> Force the app to quit by choosing Settings ⇨ Applications ⇨ Manage Installed Applications ⇨ Force Stop.

>> Clear the app's *cache* (an area of memory that the app uses to store oft-used data) by choosing Settings ⇨ Applications ⇨ Manage Installed Applications ⇨ Clear Cache.

>> Clear the app's data by choosing Settings ⇨ Applications ⇨ Manage Installed Applications ⇨ Clear Data.

>> Uninstall the app by choosing Settings ⇨ Applications ⇨ Manage Installed Applications ⇨ Uninstall, and then reinstall the app.

Chapter **12**

Ten Ways to Enhance Privacy and Security

You may think there's something, well, *odd* about this chapter. After all, isn't it a tad *weird* to be talking about Fire TV privacy and security when all you're mostly doing with Fire TV is watching TV shows and movies? Does binge-watching *The Marvelous Mrs. Maisel* require privacy? Do you have to ratchet up your security settings just to view *Despicable Me 2*?

You'd be surprised. No, watching stuff on Fire TV doesn't require enhanced privacy and security settings, but there are plenty of people out there — take a bow, nosy advertisers! — who *really* want to know what you're watching. With Fire TV, "privacy" means keeping a lid on who's tracking what you watch.

Similarly, your Fire TV device is connected to the Internet via your home network, and there are plenty of people out there — raise your hands, malicious hackers! — who *really* want to wreak havoc with your data. With Fire TV, "security" means taking steps to ensure that online miscreants don't get a foot in your digital door.

In this chapter, you investigate ten ways to enhance the privacy and security of your Fire TV device and your Amazon account. Sure, it takes a bit of time to implement these measures, but believe me when I tell you that it'll be worth every second.

Make Sure Your Wi-Fi Network Is Locked Up Tight

The first step in securing Fire TV is securing the network that Fire TV uses to access the Internet: your home Wi-Fi network. A secure Wi-Fi network is necessary because of a practice called *wardriving*, where a black-hat hacker drives through various neighborhoods with a portable computer or another device set up to look for available wireless networks. If the miscreant finds an unsecured network, he uses it for free Internet access (such a person is called a *piggybacker*) or to cause mischief with shared network resources.

The problem is that wireless networks are inherently unsecure because the wireless connection that enables you to access Fire TV from the kitchen or the living room can also enable an intruder from outside your home to access the network. Fortunately, you can secure your wireless network against these threats with a few tweaks and techniques:

REMEMBER

Most of what follows here requires access to your Wi-Fi router's administration or setup pages. See your router's documentation to learn how to perform these tasks.

>> **Change the router's administrator password.** By far the most important configuration chore for any new Wi-Fi router is to change the default password (and username, if your router requires one). Note that I'm talking here about the administrative password, which is the password you use to log on to the router's setup pages. This password has nothing to do with the password you use to log on to your Internet service provider (ISP) or to your wireless network. Changing the default administrative password is crucial because it's fairly easy for a nearby malicious hacker to access your router's login page and all new routers use common (and, therefore, well-known) default passwords (such as "password") and usernames (such as "admin").

>> **Change the Wi-Fi network password.** Make sure your Wi-Fi network is protected by a robust, hard-to-guess password to avoid unauthorized access. (See the sidebar "Coming up with a strong password," a bit later in this chapter.)

>> **Beef up your Wi-Fi router's encryption.** To ensure that no nearby mischief-maker can intercept your network data (using a tool called a *packet sniffer*), you need to encrypt your wireless network. Some older routers either have no encryption turned on or use an outdated (*read:* unsecure) encryption called Wired Equivalent Privacy (WEP). The current gold standard for encryption is Wi-Fi Protected Access II (WPA2), so make sure your router uses this security type.

COMING UP WITH A STRONG PASSWORD

As I show in this chapter, making Fire TV more secure involves setting passwords on three things: your Wi-Fi network and your Wi-Fi router's administration app, which I talk about in "Make Sure Your Wi-Fi Network Is Locked Up Tight," as well as your Amazon account, which I discuss later (see "Secure Your Amazon Account with a Strong Password"). However, it's not enough to just use any old password that pops into your head. To ensure the strongest security for your Fire TV system, you need to make each password robust enough that it's impossible to guess and impervious to software programs designed to try different password combinations. Such a password is called a *strong password*. Ideally, you should build a password that provides maximum protection while still being easy to remember.

Lots of books will suggest ridiculously abstruse password schemes (I've written some of those books myself), but you really need to know only three things to create strong-like-a-bull passwords:

- **Use passwords that are at least 12 characters long.** Shorter passwords are susceptible to programs that just try every letter combination. You can combine the 26 letters of the alphabet into about 12 million five-letter word combinations, which is no big deal for a fast program. If you use 12-letter passwords — as many experts recommend — the number of combinations goes beyond mind-boggling: 90 quadrillion, or 90,000 trillion!

- **Mix up your character types.** The secret to a strong password is to include characters from the following categories: lowercase letters, uppercase letters, punctuation marks, numbers, and symbols. If you include at least one character from three (or, even better, all five) of these categories, you're well on your way to a strong password.

- **Don't be obvious.** Because forgetting a password is inconvenient, many people use meaningful words or numbers so that their passwords will be easier to remember. Unfortunately, this means that they often use extremely obvious things such as their name, the name of a family member or colleague, their birth date, or their Social Security number. Being this obvious is just asking for trouble. Adding 123 or ! to the end of the password doesn't help much either. Password-cracking programs try those.

>> **Check your network name for identifying info.** Make sure the name of your Wi-Fi network — known as its *service set identifier* (SSID) — doesn't include any text that identifies you (for example, "Joe Flaherty's Network") or your location ("123 Primrose Lane Wi-Fi").

>> **Update your router's firmware.** The internal program that runs the Wi-Fi router is called its *firmware*. Reputable router manufacturers release regular firmware updates not only to fix problems and provide new features, but also to plug security holes. Therefore, it's crucial to always keep your router's firmware up to date.

Stop Saving Wi-Fi Passwords to Amazon

Fire TV devices, Fire TV Edition devices, Alexa-enabled devices, and smart-home devices all require access to your Wi-Fi network to do their jobs. That's reasonable, but what's unreasonable is that every one of those devices won't work until you log each one in to your network (which, of course, is protected by a strong password, as I describe in the "Make Sure Your Wi-Fi Network is Locked Up Tight" section of this chapter). That's a pain, but Amazon is trying to reduce that bother by offering a feature called Wi-Fi Simple Setup, which requires two things:

>> **A device compatible with Wi-Fi Simple Setup that's already connected to your Wi-Fi network.**

REMEMBER

Fire TV devices compatible with Wi-Fi Simple Setup are the Fire TV Stick (second generation and later), Fire TV Stick 4K, and Fire TV Cube. Other Amazon devices that support Wi-Fi Simple Setup are the second-generation or later Echo, the second-generation or later Echo Dot, and all generations of the Echo Plus, Echo Show, Echo Show 5, Amazon Smart Plug, and AmazonBasics Microwave.

>> **The password to your Wi-Fi network saved to Amazon.**

If you've checked off both items, then setting up a new device that's compatible with Wi-Fi Simple Setup is ridiculously easy. If you purchase your Wi-Fi Simple Setup device from Amazon, then Amazon automatically associates the device with your Amazon account, which means that when you plug in the device, it will connect to your network automatically using your saved Wi-Fi password. Now *that's* easy!

In Chapter 3, I mention that one step in the Fire TV setup asks if you want to save your Wi-Fi password to Amazon. I encourage you to choose Yes not only to make it easier to set up other Wi-Fi Simple Setup devices, but also because Amazon has gone to great lengths to ensure that your saved network password is safe:

>> Amazon's Privacy Policy states that it will not share your Wi-Fi password with a third party without your permission.

>> The password is stored in encrypted form on the server.

>> Devices that ask for network access are first authenticated by Amazon.

>> When needed, the password is sent using an encrypted connection.

These security steps are reassuring, but you may still feel more than a little uneasy having the password to your home network stored in the cloud. And, yes, Amazon authenticates third-party devices that want on your network, but can you really be sure that no rogue device can also breach your network?

To allay these justifiable fears, you can turn off Wi-Fi Simple Setup and delete your saved passwords using either Fire TV or the Amazon website.

Delete saved Wi-Fi passwords via Fire TV

Follow these steps on your Fire TV device to stop saving your Wi-Fi passwords and delete any saved passwords:

1. **On your Fire TV device, choose Settings ⇨ Network.**

The Network settings appear. Scroll to the bottom of this screen to see the Save Wi-Fi Passwords to Amazon setting, which will likely be set to On, as shown in Figure 12-1.

FIGURE 12-1: Use the Save Wi-Fi Passwords to Amazon setting to remove your saved passwords from Amazon.

2. **Choose Save Wi-Fi Passwords to Amazon.**

Fire TV opens the Delete Saved Wi-Fi Passwords screen, which asks you to confirm not only that you no longer want to save your network password, but also that you want to delete any Wi-Fi passwords that are currently saved to Amazon.

3. **Choose Delete.**

   Fire TV stops saving your Wi-Fi passwords and removes any saved passwords from your Amazon account.

Delete saved Wi-Fi passwords via Amazon

If you don't have access to your Fire TV device, you can still turn off Wi-Fi Simple Setup and delete your saved Wi-Fi passwords directly from the Amazon website. Here are the steps to follow:

1. **Surf to www.amazon.com (or your country's Amazon domain) and sign in to your account.**

2. **Choose Account & Lists ⇨ Your Content and Devices.**

   On other Amazon domains, you may have to choose Account & Lists ⇨ Manage Your Content and Devices.

3. **Choose the Preferences tab.**

4. **Choose Saved Wi-Fi Passwords.**

   The Saved Wi-Fi Passwords settings appear, as shown in Figure 12-2.

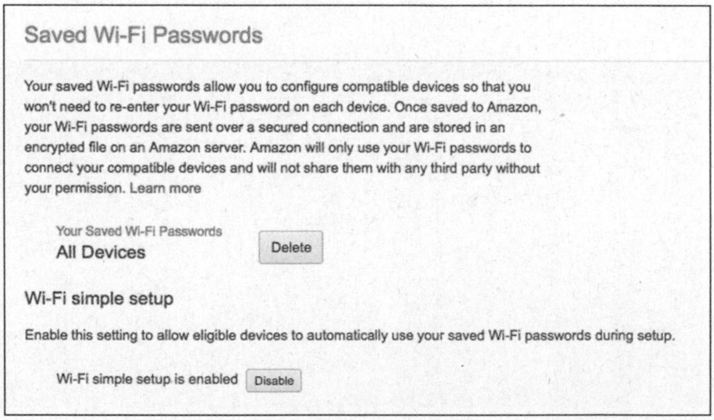

Saved Wi-Fi Passwords

Your saved Wi-Fi passwords allow you to configure compatible devices so that you won't need to re-enter your Wi-Fi password on each device. Once saved to Amazon, your Wi-Fi passwords are sent over a secured connection and are stored in an encrypted file on an Amazon server. Amazon will only use your Wi-Fi passwords to connect your compatible devices and will not share them with any third party without your permission. Learn more

Your Saved Wi-Fi Passwords
All Devices   [Delete]

Wi-Fi simple setup

Enable this setting to allow eligible devices to automatically use your saved Wi-Fi passwords during setup.

Wi-Fi simple setup is enabled [Disable]

FIGURE 12-2:
Use the Saved Wi-Fi Passwords section to remove your saved passwords from Amazon.

5. **To remove your saved network password from Amazon, choose Delete.**

   Amazon asks you to confirm the deletion.

6. **Choose Yes, Delete Permanently.**

   Amazon deletes your saved Wi-Fi password.

7. **To prevent devices from using Wi-Fi Simple Setup, choose Disable.**

   Amazon asks you to confirm.

8. **Choose Yes, Disable.**

   Amazon disables Wi-Fi Simple Setup.

Secure Your Amazon Account with a Strong Password

Because everything Fire TV does is tied to your Amazon account, your Fire TV experience is only as secure as your Amazon account. Therefore, it's vital to ensure that you've got your Amazon account locked down. Fortunately, that requires just two things: giving your account a strong password (as I describe in this section) and turning on Amazon's Two-Step Verification feature (which I discuss in the next section).

Your Amazon account's first line of defense is a strong password. First, check back in this chapter to the "Coming up with a strong password" sidebar. After you've got a bulletproof password figured out, follow these steps to change your existing Amazon password:

1. **Surf to www.amazon.com (or your country's Amazon domain) and sign in to your account.**

2. **Choose Account & Lists ⇨ Your Account.**

3. **Choose Login & Security.**

   If Amazon asks you to sign in, type your password and choose Sign-In.

4. **Choose the Edit button beside the Password setting.**

   Amazon displays the Change Password page, shown in Figure 12-3.

5. **Type your current password.**

6. **Type your new, strong password in the two text boxes.**

7. **Choose Save Changes.**

   Amazon applies the new password to your account.

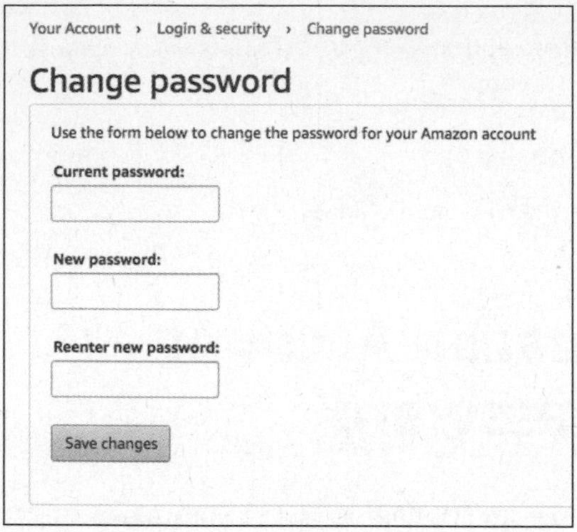

FIGURE 12-3:
Use the Change Password page to modify your Amazon account password.

Enable Amazon's Two-Step Verification

A password made of steel is a necessary security feature, but, sadly, it's not a sufficient security feature. A malicious user may still worm his way into your account with guile or brute force, so you need a second line of defense. That line is a feature that Amazon calls Two-Step Verification (which is a more comprehensible name than what the rest of the Internet most often uses for the same feature: Two-Factor Authentication; note that Amazon often abbreviates Two-Step Verification to 2SV). The "Two-Step" part means that getting access to your Amazon account requires two separate actions:

1. **Sign in using your Amazon account credentials.**

2. **Verify that you're authorized to access the account by entering a code that Amazon sends to you.**

   This code is called a *one-time password* (OTP). Amazon sends the OTP to a two-step verification *authenticator,* which can be either (or both) of the following:

   - **Your mobile phone:** In this case, Amazon sends the OTP in a text message.

   - **An authenticator app installed on your mobile phone:** In this case, the app generates the OTP. To use this method, you need to install an authenticator app on your phone. Popular authenticator apps are Google Authenticator, Microsoft Authenticator, and Duo Mobile.

Amazon requires *two* authenticators, so most people use a mobile phone number and an authenticator app.

Here are the steps to follow to enable Two-Step Verification and tell Amazon how you want to receive your verification codes:

1. **Surf to www.amazon.com (or your country's Amazon domain) and sign in to your account.**

2. **Choose Account & Lists ➪ Your Account.**

3. **Choose Login & Security.**

4. **Choose the Edit button beside Advanced Security Settings.**

   Amazon displays the Advanced Security Settings page.

5. **Choose the Get Started button that appears to the right of the Two-Step Verification label.**

   Amazon asks how you want to receive your Two-Step Verification codes by asking you to select which type of authenticator you want to use, as shown in Figure 12-4.

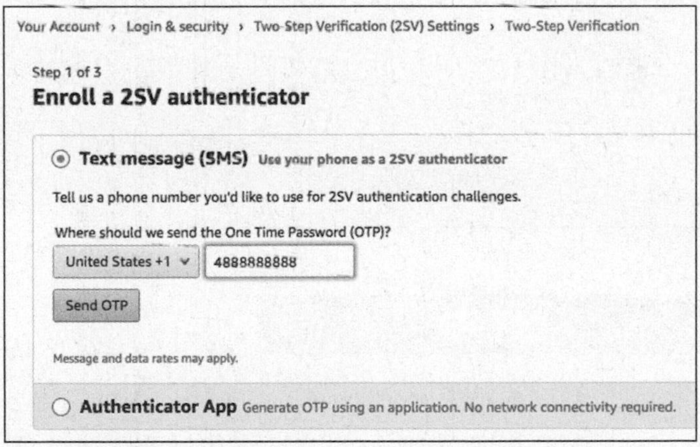

FIGURE 12-4: Use this page to specify how you want to receive your Two-Step Verification codes.

6. **Select the Text Message radio button.**

7. **Enter your mobile phone number, and then choose Send OTP.**

   Amazon sends you a test OTP in a text message.

8. Use the Enter the One Time Password (OTP) text box to type the code you received, and then choose the Verify OTP and Continue button.

Amazon now asks you to select a second verification method. The following steps assume you're using an authenticator app, so be sure to have an app installed before continuing.

9. Select the Authenticator App radio button.

10. Open your authenticator app and use the app to scan the barcode displayed by Amazon, as shown in Figure 12-5.

The authenticator app adds your Amazon account and displays a one-time password, as shown in Figure 12-6.

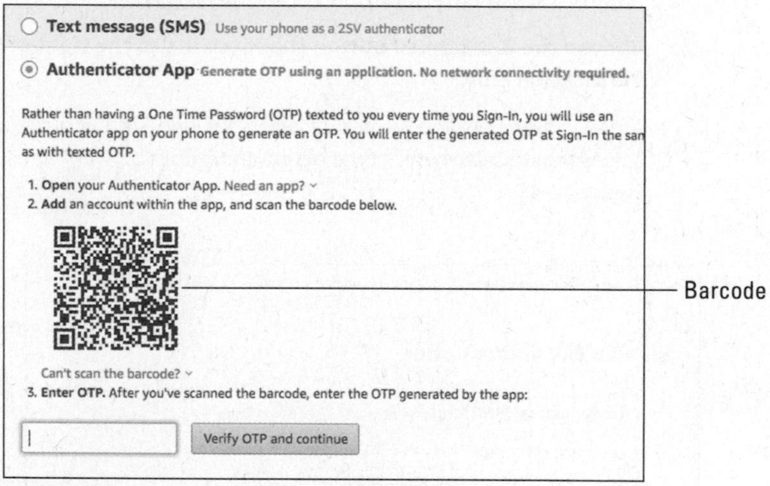

○ **Text message (SMS)** Use your phone as a 2SV authenticator

◉ **Authenticator App** Generate OTP using an application. No network connectivity required.

Rather than having a One Time Password (OTP) texted to you every time you Sign-In, you will use an Authenticator app on your phone to generate an OTP. You will enter the generated OTP at Sign-In the sam as with texted OTP.

1. **Open** your Authenticator App. Need an app? ⌄
2. **Add** an account within the app, and scan the barcode below.

[QR code] —————————— Barcode

Can't scan the barcode? ⌄
3. **Enter OTP.** After you've scanned the barcode, enter the OTP generated by the app:

[text box] [Verify OTP and continue]

FIGURE 12-5:
Use your authenticator app to scan the displayed barcode.

Edit   DUO   + ≡

THIRD-PARTY
IFTTT   ⌄

a   AMAZON
mail@mcfedries.com   ⌃

848 448   ⑲

FIGURE 12-6:
The authenticator app adds your Amazon account and generates an OTP.

11. **Return to the Amazon web page and use the Enter OTP text box to type the OTP as it's displayed in the app (see Figure 12-7), and then choose the Verify OTP and Continue button.**

When entering the OTP, don't dawdle because the app only gives you a short time — usually a minute or less — to enter the code before it generates a new one.

Amazon now displays some information about using Two-Step Verification on devices that can't display a second screen to enter the verification code.

3. **Enter OTP.** After you've scanned the barcode, enter the OTP generated by the app:

848448   [Verify OTP and continue]

12. **Choose Got It. Turn on Two-Step Verification.**

Two-Step Verification is now active on your Amazon account.

Prevent Fire TV from Playing Previews Automatically

One of the more annoying habits of Fire TV rears its head when you navigate to the Movies or TV Shows screen, both of which display the Featured carousel at the top, which shows five movies or TV series that Amazon is highlighting. The problem occurs when you press the Fire TV remote's Down button once, which takes you into the Featured carousel. If you don't press Down again very quickly, Fire TV automatically starts playing the first Featured preview. That's not only grimace-inducing, but also a minor privacy violation because someone sitting nearby may believe you're watching the latest *SpongeBob SquarePants* show.

To prevent Fire TV from automatically playing Featured previews, follow these steps:

1. **On your Fire TV device, choose Settings ⇨ Preferences ⇨ Featured Content.**

The Featured Content screen appears.

2. **To prevent preview videos from playing automatically, choose Allow Video Autoplay to set that option to Off.**

   When you scroll into the Featured carousel, Fire TV still rotates through the featured items, but now it only shows a still image for each item rather than a video preview.

3. **To prevent just the audio portion of the previews from playing automatically, choose Allow Audio Autoplay to set that option to Off.**

   Fire TV stops automatically playing preview video and/or audio.

Remove Content from Your Fire TV

Fire TV only gives you a limited number of options for removing content. For example, if you don't guests or want anyone in your household to know what you've watched recently, you can remove some or all of the items that appear in the Home screen's Recent list. Here are the content types you can remove from Fire TV:

>> **Apps:** Choose Settings ➪ Applications ➪ Manage Installed Applications. Choose the app you want to remove and then choose Uninstall. When Fire TV asks you to confirm, choose Uninstall.

>> **Recent items:** On the Fire TV Home screen, select an item in the Recent row, press the Fire TV remote's Menu button, and then choose Remove from Recent.

>> **Watchlist items:** In the Fire TV Your Videos screen, select an item in the Watchlist row, press the Fire TV remote's Menu button, and then choose Remove from Watchlist.

Prevent Amazon from Using Personal Data for Marketing

As you use Fire TV, the device keeps track of your choices, preferences, and other personal data, which Amazon uses both to serve you more relevant ads and to help improve future versions of Fire OS (the operating system that runs Fire TV). Collecting your personal data in this way isn't necessarily evil, but you may not be comfortable with having your device tracking your every move in the Fire TV interface. If so, go ahead and follow these steps to turn off this tracking:

1. **On your Fire TV device, choose Settings ⇨ Preferences ⇨ Privacy Settings.**

   The Privacy Settings screen appears.

2. **Choose Device Usage Data.**

   Fire TV opens the Device Usage Data screen, which explains how Amazon uses your personal data and what turning off this setting means, as shown in Figure 12-8.

REMEMBER

   Turning off Device Usage Data does *not* mean that you'll see no more ads. Boo! Instead, it just means that the ads you see will likely be less relevant.

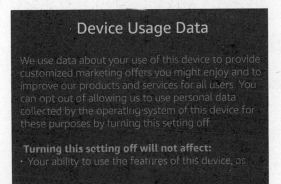

FIGURE 12-8:
The Device Usage
Data screen
explains this
setting and
enables you to
turn it off.

3. **Choose Turn Off.**

   Fire TV stops sending your personal data to Amazon.

Prevent Amazon from Tracking Your App Usage

By default, Amazon uses some of your Fire TV device–generated data to improve and enhance Fire TV. There are two types of data that Amazon uses to make Fire TV better:

>> **Third-party app usage:** Amazon tracks how often you use third-party apps and for how long you use each app.

>> **Over-the-air TV:** Amazon tracks what you watch when you tune into over-the-air programming.

If you're uncomfortable having your Fire TV usage tracked in either or both ways, you can configure your device to not include this data when it's improving Fire TV. Here are the steps to follow:

1. **On your Fire TV device, choose Settings ⇨ Preferences ⇨ Privacy Settings.**

    The Privacy Settings screen appears.

2. **Choose Collect App and Over-the-Air Usage Data.**

    Fire TV opens the Collect App and Over-the-Air Usage Data screen, which explains how Amazon tracks your third-party app and over-the-air TV usage, as shown in Figure 12-9.

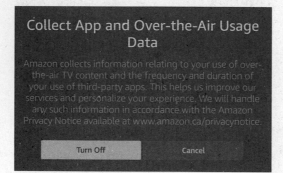

Collect App and Over-the-Air Usage Data

Amazon collects information relating to your use of over-the-air TV content and the frequency and duration of your use of third-party apps. This helps us improve our services and personalize your experience. We will handle any such information in accordance with the Amazon Privacy Notice available at www.amazon.ca/privacynotice.

Turn Off   Cancel

3. **Choose Turn Off.**

    Fire TV stops sending your third-party app and over-the-air TV usage data to Amazon.

Prevent Advertisers from Tracking You

Amazon labels your Fire TV device with a unique identifier called an *advertising ID*, which Amazon provides to ad networks that display ads within third-party apps. Why would Amazon do such a thing? The advertising ID enables ad networks to track your Fire TV device across multiple apps, which lets the network build a profile of you, from which the network can then show ads that are targeted toward you and your interests.

Has Amazon gone over to the Dark Side? No, not really. The goal here is to allow ad networks to learn enough about your Fire TV life that it can serve you ads that are relevant to your interests, rather than ads that are chosen randomly. The thinking is that if you see a targeted ad, then you're more likely to choose that ad and the developer of the app showing the ad will get a little more money. It's almost noble, in a way.

Almost. The other side of ad tracking is, well, the *tracking* part, which effectively means that ad networks are monitoring your Fire TV behavior across multiple apps to build up a profile for you. If that strikes you as positively dystopian, you're not alone. Many people dislike being tracked in this way, and web browsers such as Firefox are now shipping with tracking prevention turned on.

If the idea of ad networks monitoring what you do while you use Fire TV is off-putting and creepy, follow these steps to prevent ad networks from tracking you using your Fire TV advertising ID:

1. **On your Fire TV device, choose Settings ➪ Preferences ➪ Privacy Settings.**

The Privacy Settings screen appears.

2. **If you just want to reset your advertising ID (which forces ad networks to rebuild your profile from scratch), highlight the Your Advertising ID setting, and then press Menu on your Fire TV remote.**

3. **To turn off ad tracking, highlight the Interest-Based Ads setting (see Figure 12-10) and then press the Select button to turn that setting to Off.**

Fire TV opts you out of ad tracking.

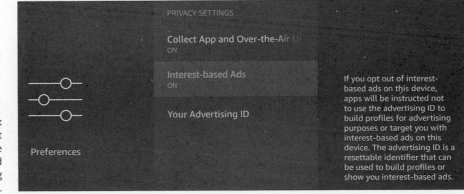

FIGURE 12-10: Press Select on the Interest-Based Ads setting to turn it off.

Deregister Your Fire TV Device

If you're selling or giving away a Fire TV device, you want to make sure the device is wiped clean of all your custom settings and personal info. You can do this by resetting your device, which I describe in Chapter 11.

You also want to ensure that the device is no longer connected to your Amazon account, and that means *deregistering* the device with Amazon. You have two choices:

» In Fire TV, choose Settings ⇨ My Account ⇨ Amazon Account ⇨ Deregister. When Fire TV asks you to confirm, choose Deregister.

» Surf to www.amazon.com (or your country's Amazon domain), sign in to your account, choose Account & Lists ⇨ Your Content and Devices (or Account & Lists ⇨ Manage Your Content and Devices), select the Devices tab, choose the Action button beside the device you're getting rid of, and then choose Deregister. When Amazon asks you to confirm, choose Deregister.

Index

About the Author

Paul McFedries is the president of Logophilia Limited, a technical writing company. He has worked with computers large and small since 1975. Although he's now primarily a writer, Paul has worked as a programmer, consultant, database developer, and website developer. He has written nearly 100 books that have sold more than four million copies throughout the galaxy. For fun, Paul bakes bread and doesn't let anyone pass him when he's running. Paul invites everyone to drop by his personal website (www.mcfedries.com) or follow him on Twitter (@paulmcf).

Dedication

To Karen and Chase, who make life beautiful.

Author's Acknowledgments

If we're ever at the same cocktail party and you overhear me saying something like "I wrote a book," I hereby give you permission to wag your finger at me and say "Tsk, tsk." Why the scolding? Because although I did write this book's text and take its screenshots, that represents only a part of what constitutes a "book." The rest of it is brought to you by the dedication and professionalism of Wiley's editorial and production teams, who toiled long and hard to turn my text and images into an actual book.

I offer my sincere gratitude to everyone at Wiley who made this book possible, but I'd like to extend an enthusiastic "Thanks a bunch!" to the folks I worked with directly: Associate Publisher Katie Mohr and Project Editor Elizabeth Kuball. I'd also like to thank my agent, Carole Jelen, for helping to make this project possible.

Publisher's Acknowledgments

Associate Publisher: Katie Mohr
Project Editor: Elizabeth Kuball
Copy Editor: Elizabeth Kuball
Sr. Editorial Assistant: Cherie Case

Production Editor: Siddique Shaik
Cover Image: Courtesy of Amazon.com